Christiane Piegholdt

Effects of tillage intensity on soil C, N, and P pools

Christiane Piegholdt

Effects of tillage intensity on soil C, N, and P pools

Pools with different stability

Südwestdeutscher Verlag für Hochschulschriften

Impressum / Imprint

Bibliografische Information der Deutschen Nationalbibliothek: Die Deutsche Nationalbibliothek verzeichnet diese Publikation in der Deutschen Nationalbibliografie; detaillierte bibliografische Daten sind im Internet über http://dnb.d-nb.de abrufbar.
Alle in diesem Buch genannten Marken und Produktnamen unterliegen warenzeichen-, marken- oder patentrechtlichem Schutz bzw. sind Warenzeichen oder eingetragene Warenzeichen der jeweiligen Inhaber. Die Wiedergabe von Marken, Produktnamen, Gebrauchsnamen, Handelsnamen, Warenbezeichnungen u.s.w. in diesem Werk berechtigt auch ohne besondere Kennzeichnung nicht zu der Annahme, dass solche Namen im Sinne der Warenzeichen- und Markenschutzgesetzgebung als frei zu betrachten wären und daher von jedermann benutzt werden dürften.

Bibliographic information published by the Deutsche Nationalbibliothek: The Deutsche Nationalbibliothek lists this publication in the Deutsche Nationalbibliografie; detailed bibliographic data are available in the Internet at http://dnb.d-nb.de.
Any brand names and product names mentioned in this book are subject to trademark, brand or patent protection and are trademarks or registered trademarks of their respective holders. The use of brand names, product names, common names, trade names, product descriptions etc. even without a particular marking in this work is in no way to be construed to mean that such names may be regarded as unrestricted in respect of trademark and brand protection legislation and could thus be used by anyone.

Coverbild / Cover image: www.ingimage.com

Verlag / Publisher:
Südwestdeutscher Verlag für Hochschulschriften
ist ein Imprint der / is a trademark of
OmniScriptum GmbH & Co. KG
Heinrich-Böcking-Str. 6-8, 66121 Saarbrücken, Deutschland / Germany
Email: info@svh-verlag.de

Herstellung: siehe letzte Seite /
Printed at: see last page
ISBN: 978-3-8381-3964-7

Zugl. / Approved by: Witzenhausen, Universität Kassel, Diss., 2013

Copyright © 2015 OmniScriptum GmbH & Co. KG
Alle Rechte vorbehalten. / All rights reserved. Saarbrücken 2015

This work has been accepted by the Faculty of Organic Agricultural Sciences of the University of Kassel as a thesis for acquiring the academic degree of Doktor der Naturwissenschaften (Dr. rer. nat.).

First supervisor: Prof. Dr. Bernard Ludwig
Second supervisor: Prof. Dr. Rainer Georg Jörgensen

Acknowledgments

First I sincerely thank Prof. Dr. Bernard Ludwig, my supervisor and the spokesman of the DFG Research Training Group. With his continuous support and encouragement he gave me the opportunity to prepare this dissertation.

Furthermore, I want to thank Prof. Dr. Rainer Georg Jörgensen for co-supervising this work.

I am grateful to Dr. Daniel Geisseler and Dr. Michael Kaiser for their scientific guidance.

I thank Dr. Anna Jacobs, Dr. Felix Heitkamp, and Dr. Anja Sänger, my former colleagues of the DFG Research Training Group, as well as Dr. Shafique Maqsood for receiving me here and the warm welcome in the Department of Environmental Chemistry, and for helpful advices, even after their disputation and leaving our department to new challenges.

Thanks to Rouven Andruschkewitsch, Deborah Linsler, and Johanna Pinggera, my colleagues in the Department of Environmental Chemistry, for co-operation within experiments, laboratory work, as well as interpretation and discussion of data during our common years as doctoral candidates.

Lot of thanks thank Anja Sawallisch for the continuous support in the laboratory, and I thank Dr. Christel Ross for the great organization of the Research Training Group. I am grateful to both of them for the motivation during my doctoral thesis.

I thank the Department of Soil Biology and Plant Nutrition, especially Gabi Dormann, and the Department of Organic Farming and Cropping Systems for the opportunity to use their equipment.

For the opportunity to prepare this dissertation I thank the German Research Foundation, which finances the Research Training Group 1397 "Regulation of soil organic matter and nutrient turnover in organic agriculture".

Preface

This thesis was prepared within the Research Training Group 1397 "Regulation of soil organic matter and nutrient turnover in organic agriculture" and funded by a grant of the German Research Foundation (DFG). The thesis is submitted to the Faculty of Organic Agricultural Sciences to fulfill the requirements for the degree "Doktor der Naturwissenschaften" (Dr. rer. Nat.).

The cumulative dissertation is based on three papers as first author or co-author, which are published, submitted, or in preparation for submission to international refereed journals. The manuscripts are included in chapters 2, 3 and 4.

The general introduction (chapter 1) focus on theoretical and methodological issues, specific introductions on the effect of tillage treatments and/or litter distribution on soil are given in the separate manuscripts (chapters 2, 3 and 4). Following the manuscripts, a general conclusion as a synthesis is given in chapter 5.

This thesis includes the following papers:

Chapter 2:

Kaiser, M.; Piegholdt, C.; Andruschkewitsch, R.; Koch, H.-J. and Ludwig, B.: Impact of tillage intensity on carbon and nitrogen pools in surface and subsurface soils of four long-term field experiments. Eur. J. Soil Sci., in preparation for submission.

Chapter 3:

Piegholdt, C.; Geisseler, D.; Koch, H.-J. and Ludwig, B.: Long-term tillage effects on the distribution of P fractions of German loess soils. J. Plant Nutr. Soil Sci., accepted.

Chapter 4:

Piegholdt, C.; Linsler, D.; Andruschkewitsch, R. and Ludwig, B.: Effects of different winter cover crops and residue location on C and N dynamics in an incubation experiment. In preparation for submission.

Table of contents

List of tables II

List of figures V

List of abbreviations VII

Summary VIII

Zusammenfassung XI

1 General introduction 1

    1.1 Impact of tillage intensity on OM transformation 2

    1.2 Drivers for stabilization and destabilization of SOM and soil P 3

    1.3 SOM pools with different turnover times 4

    1.4 Soil P pools with different stability 6

    1.5 General objectives 7

    1.6 References 8

2 Impact of tillage intensity on carbon and nitrogen pools in surface and subsurface soils of four long-term field experiments 11

    2.1 Introduction 12

    2.2 Materials and methods 16

    2.3 Results 22

    2.4 Discussion 32

    2.5 Conclusions 38

    2.6 References 39

3 Long-term tillage effects on the distribution of Phosphorus fractions in German loess soils 44

    3.1 Introduction 45

    3.2 Material and methods 47

    3.3 Results and discussion 52

    3.4 Conclusions 64

    3.5 References 64

4 Effects of different winter cover crops and residue location on C and N dynamics in an incubation experiment 68

    4.1 Introduction 69

    4.2 Materials and methods 70

    4.3 Results and discussion 77

    4.4 Conclusions 83

    4.5 References 83

5 General conclusions 87

List of tables

Table 1: Site designations, year of establishment, altitude, and climatic conditions of the long-term field experiments and mean values of the soil sand, silt, and clay, as well as oxalate soluble Al and Fe (Al_{ox}, Fe_{ox}) stocks (0- ~25 cm) and the pH values. The soil data shown are mean values of three pseudo replicates and the standard errors are given in parentheses. 16

Table 2: Site, tillage system, soil depth, and the respective stocks of organic C (C_{org}), and total N (N_t) as well as of the labile, intermediate, and passive C and N pools, the microbial biomass C and N (C_{mic}, N_{mic}), and the mineralized C (CO_2-C) and N (N_{min}). The data shown for passive C and N, C_{mic} and N_{mic} are mean values of three pseudo replicates. The data shown for labile C and N, intermediate C and N, CO_2-C, and N_{min} are mean values of the two lab replicates (three pseudo replicates per site, treatment and depth were mixed for each lab replicate). Values in parentheses are standard errors. 23

Table 3: Tillage system, soil depth, and the respective stocks of the soil organic C (C_{org}), total N (N_t), the light (LF) and heavy fraction (HF), the microbial biomass C and N (C_{mic}, N_{mic}). The data shown are mean values of the four study sites, the standard errors are given in parentheses. Values followed by different lower case letters are significantly different ($p \leq 0.05$) and values followed by upper case letters are different by trend ($p \leq 0.1$). Letters refer to the comparison of tillage treatments within one depth. 24

Table 4: Climatic conditions (long-term annual means) and soil properties at the four research sites from the long-term field experiment of the Institute for Sugar Beet Research and the Agricultural Division of the Südzucker AG. Values shown for texture are means of the three pseudoreplicates per tillage system and study site. 48

Table 5: Crop yields and Calcium-Acetate-Lactate extractable P (CAL-P) status at the experimental sites under conventional tillage (CT) and no-till (NT). Values shown are means and standard errors of the mean (winter wheat n = 5, sugar beet n = 2 (means for 5 and 2 years, respectively), CAL-P status n = 3 (means for 3 pseudoreplicates for each field)). 53

List of tables

Table 6: Contents of organic and carbonate-C (C_{org}, CO_3-C), total N (N_t), oxalat soluble Al and Fe (Al_{ox}, Fe_{ox}), as well as pH in the soil profile under conventional tillage (CT) and no-till (NT). Values shown are for means and standard errors for the individual sites (n = 3 pseudoreplicates) and means and standard errors of the sites (n = 4 field replicates). Values followed by the same letter are not significantly different (p ≤0.05). Capital letters refer to the comparison of tillage treatments within one depth, while lower case letters refer to sampling depth within one tillage treatment. 54

Table 7: Means, standard deviations and significance values of P-fractions and P stocks (inorganic P: P_i, organic P: P_o) of both tillage treatments (conventional tillage: CT, no-till: NT) and depths (n = 4). Values followed by the same letter are not significantly different (P <0.05). Capital letters refer to the comparison of tillage treatments within one depth, while lower case letters refer to sampling depth within one tillage treatment. 58

Table 8: Stepwise multiple regression analysis of P fractions (expressed in mg kg^{-1} soil) on soil properties, including clay content (clay; g kg^{-1} soil), oxalate-soluble Fe and Al (Fe_{ox} and Al_{ox} (mg kg^{-1} soil)), organic C (C_{org} g kg^{-1} soil), carbonate-C (CO_3-C; g kg^{-1} soil) and pH (n = 72). Unit of intercepts is mg kg^{-1} dry soil. 60

Table 9: Phosphorus contents in the fractions Calcium-Acetate-Lactate soluble P (CAL-P), water-soluble P (DI-P), and bicarbonate soluble inorganic and organic P ($NaHCO_3$-P_i and $NaHCO_3$-P_o) for each site and tillage treatment (conventional tillage: CT; no-till: NT). Values shown are means and standard errors of the means of the pseudoreplicates (n = 3). 62

Table 10: Stocks and C/N ratios, as well as C and N input of aboveground and belowground biomass for soil samples of 0-25 cm depth. The data shown are mean values of the three column replicates and the standard errors of the means are given in parentheses (n = 3). Values followed by different letters are significantly different (p ≤0.05). Letters refer to the comparison of cover crop species. 77

List of tables

Table 11: Stocks of organic C (C_{org}), total N (N_t), and microbial C and N (C_{mic}, N_{mic}) of soil samples from 0-25 cm depth of the three cover crop and the fallow after the growth of the crops (T1). The data shown are mean values of the three column replicates and the standard errors are given in parentheses (n = 3). 78

Table 12: Stocks of leached dissolved organic carbon (DOC), mineralized N (N_{min}), and microbial C and N (C_{mic}, N_{mic}) of soil samples from 0-25 cm depth of the three cover crop and the fallow treatment for the two residue locations after freezing and within the second incubation period for 80 days (T2). The data shown are mean values of the three column replicates and the standard errors are given in parentheses (n = 3). Values followed by different letters are significantly different (p ≤0.05). Capital letters refer to the comparison of residue location within one cover crop treatment. Lower case letters refer to the comparison of cover crop treatments within one treatment of residue location. 81

List of figures

Figure 1: Cumulated stocks of emitted CO_2-C and mineralized net N (N_{min}) of soils in 0- ~5 cm as well as 5- ~25 cm depth of the three tillage treatments (CT: conventional tillage, RT: reduced tillage, NT: no-till). Datapoints are means of the four study sites, error bars refer to standard errors of the means, and the solid lines refer to the interpolated data points with an exponential one-pool model. Values followed by different lower case letters are significantly different (p $\leq$0.05) and values followed by upper case letters are different by trend (p $\leq$0.1).   26

Figure 2: Stocks of the labile, intermediate, and passive C and N pools of soils in 0- ~5 cm and 5- ~25 cm depth of the three tillage treatments (CT: conventional tillage; RT: reduced tillage; NT: no-till). Columns show the means values of the four study sites, error bars refer to standard errors of the means. Values followed by different lower case letters are significantly different (p $\leq$0.05) and values followed by upper case letters are different by trend (p $\leq$0.1). Letters refer to the comparison of tillage treatments within on depth.   29

Figure 3: Stocks of the labile C pool (all soils) versus stocks of the (a) light fraction (LF), (b) mineralized N (N_{min}), (c) microbial biomass C (C_{mic}), and (d) microbial biomass N (N_{mic}), stocks of the intermediate C pool (all soils) versus stocks of the (e) LF, (f) N_{min}, and (g) C_{mic}, and stocks of the passive C pool (all soils) versus stocks of the (h) oxalate soluble Al (Al_{ox}), and (i) clay fraction. Correlations were analyzed with Spearman's rank correlation test.   30

Figure 4: Stocks of the labile N pool (all soils) versus stocks of the (a) light fraction (LF), (b) CO_2-C, (c) microbial biomass C (C_{mic}), and (d) microbial biomass N (N_{mic}), stocks of the intermediate N pool (all soils) versus stocks of the (e) LF, (f) N_{min} and (g) C_{mic}, and stocks the passive N pool (all soils) versus stocks of the (h) oxalate soluble Al (Al_{ox}), and (i) clay fraction. Correlations were analyzed with Spearman's rank correlation test.   31

List of figures

Figure 5: Labile (DI-P and $NaHCO_3$-P_t), stable (NaOH-P_t and HCl-P) and residual-P content in the soil profile under conventional tillage (CT) and no-till (NT). Error bars refer to standard error (n = 4). 61

Figure 6: Scheme of the experimental design, sampling times and analyses carried out respectively for the three cover crop and the fallow treatment. 73

Figure 7: Cumulative amounts of CO_2-C (t ha^{-1}), N_2O-N (kg ha^{-1}), and mineralized net N (kg ha^{-1}) of soil samples from 0-25 cm depth of the three cover crop and the fallow treatment for the two residue locations after freezing and within the second incubation period for 80 days (T2). Datapoints are means of the three column replicates, error bars refer to standard errors of the means (n = 3). Values followed by different letters are significantly different (p ≤0.05). Letters refer to the comparison of cover crop treatments within one treatment of residue location. 80

Figure 8: Stocks of labile C (based on cumulative CO_2-C emission) of soil samples from 0-25 cm depth of the three cover crop and the fallow treatment for the two residue locations after freezing and within the second incubation period for 80 days (T2). Columns show the mean values of the three column replicates, error bars refer to standard errors of the means (n = 3). Values followed by different letters are significantly different (p ≤0.05). Letters refer to the comparison of cover crop treatments within one treatment of residue location. 82

List of abbreviations

Al_{ox}, Fe_{ox}	Oxalate-soluble aluminum and iron
C_{mic}	Microbial biomass carbon
C_{org}	Soil organic carbon
CT	Conventional tillage
DOC	Dissolved organic carbon
HF	Heavy fraction ($\rho > 1.8$ g cm^{-3})
LF	Light fraction ($\rho \leq 1.8$ g cm^{-3})
LFE	Long-term field experiment
MRT	Mean residence time
N_{mic}	Microbial biomass nitrogen
N_{min}	Mineralized nitrogen
Nt	Total nitrogen
NT	No-till
OM	Organic matter
P	Phosphorus
P_i, P_o, P_t	Inorganic, organic and total Phosphorus
RT	Reduced tillage
SOM	Soil organic matter
WHC	Water holding capacity

Summary

The dynamics of soil organic matter turnover are known to be affected by tillage intensity due to different distribution of crop residues, different aggregate dynamics, and erosion losses. Until now, only little data about the influence of tillage intensity and cover crop cultivation on C, N, and P dynamics in the surface (0-5 cm soil depth) and subsurface layer (5-25 cm soil depth) of loess soils are available. The main objectives of this work were to:

(i) quantify the influence of different long-term tillage systems on labile, intermediate, and passive C and N pools by physical and chemical as well as by mathematical investigation methods;

(ii) quantify the influence of different long-term tillage systems on P fractions with different availability for plants uptake by using a stepwise chemical fractionation procedure to divide P into pools with different stability in soil;

(iii) quantify the impact of the integration of cover crops in a crop rotation in combination with a different distribution of the plant residues in the soil on mineralizable C and N.

The results of the first and second sub-experiment based on investigations on four long-term field experiments (LFEs) in east and south Germany. They were initiated between 1990 and 1997 by the Institute of Sugar Beet Research (IfZ) in cooperation with the Südzucker AG. Each LFE comprised three tillage systems. Conventional tillage (CT) included mouldboard ploughing to a depth of 25-30 cm, reduced tillage (RT) included cultivation down to 15 cm, and the no-till system (NT) without any soil tillage, except a seedbed preparation down to 5 cm before sugar beets are sown. The results of the third part of this study thesis based on an incubation experiment, which investigated the decomposition dynamic of different distributed plant residues by measurement of mineralized C and N under controlled conditions.

Summary

According to the main objectives, the following research results were detected:

(i) Higher stocks of labile C and N pools were found in the surface soil under NT (C: 1.76 t ha^{-1}, N: 166 kg ha^{-1}), compared to CT (C: 0.44 t ha^{-1}, N: 52 kg ha^{-1}). In contrast, the labile C stocks in subsurface soils were higher under CT with 2.68 t ha^{-1} compared to NT with 2 t ha^{-1} and RT with 1.87 t ha^{-1}. Stocks of intermediate C were accounted for 73 to 85% of total organic C, stocks of intermediate N were accounted for 70 to 95% of total N in surface and subsurface soils and so were multiple higher than the labile and passive C and N stocks. An effect of tillage intensity on the intermediate N pool was found only for the surface soil, with higher stocks under NT compared to CT. The passive C and N pools were strongly related to the determined soil mineral characteristics and independent from tillage system. In summary we found that 14 to 22 years of continuous no-till leads to higher stocks of labile C and N pools only in surface soils compared to conventional tillage and suggested a depth dependent intensity of soil organic matter dynamics.

(ii) The concentration of total P (P$_t$) in surface soil was higher in the NT treatment (792 mg kg^{-1}) and about 15% higher than the P$_t$ concentration in CT (691 mg kg^{-1}). The decrease of P$_t$ concentration with increased soil depth was higher in NT than in CT soils. This was also true for the separated P fractions, except the most stable P fraction (residual-P). In general, the tillage system had only a small effect on P concentration with higher P$_t$ concentrations in soils under NT compared to CT. This resulted presumably from the shallower incorporation of plant residues compared to CT.

(iii) In the cover crop experiment, the biomass growth was highest for mustard and decreased in the order (aboveground yields in t / ha): mustard (7.0 t ha^{-1}) > phacelia (5.7 t ha^{-1}) > oil radish (4.4 t ha^{-1}). Thus, the potential of mineralizable C and N was highest in mustard covered soils. Cumulative CO_2 and N_2O emissions during the incubation period at 10 °C were not significantly different between the cover crop treatments and independent from the distribution of the crop residues in the soil. The stocks of cumulative leached mineralized N (N_{min}) was highest in the control soils without cover crops. The stocks of N_{min} were 51 to 72% lower in the cover crop treatments with incorporated crop residues compared to the control. At surface-applied crop residues, the stocks of N_{min} were 36 to 55% lower compared to the control. This indicated a strong benefit of cover crops on reducing nitrate-N losses by leaching between fall and spring.

Overall, reduced tillage intensity entailed in a sequestration of C and N in the soil, and the integration of cover crops in a crop rotation resulted in decreased N losses. The P availability was higher under no-till compared to conventional tillage. These findings resulted mainly from the higher concentration of organic matter in soils under reduced tillage systems compared to conventional tillage, as well as with cover crop cultivation compared to fallow soil. The results showed clearly the potential of reduced tillage intensity for the intermediate sequestration of C and N in soil, to reduce emissions of climate relevant greenhouse gases. At the same time, the concentration of plant-available P tends to increase with reduced tillage. Cover crops also increase the stocks of soil C and N, obviously independent from the cover crop species.

Zusammenfassung

Im Rahmen einer nachhaltigen Lebensmittelerzeugung gilt ein Interesse der ökologischen Landwirtschaft der Stabilisierung und Erhöhung der organischen Substanz (OS) und des Nährstoffgehalts in Böden. So steht die Produktivität von Böden im Fokus, um vergleichbare Erträge zu konventioneller Landwirtschaft zu erhalten oder diese zu erhöhen. Ein weiterer Aspekt ist mit Blick auf die globale Klimaentwicklung die Festsetzung von OS und damit von Kohlenstoff im Boden (*Franzluebbers*, 2005). Die OS und der Nährstoffgehalt in Ackerböden wird von der Bewirtschaftungsart beeinflusst, z.b. Bodenbearbeitung, Düngung und Fruchtfolge (*Seiter & Horwath*, 2004). Da OS aus einer Vielzahl verschiedener Verbindungen besteht, die alle unterschiedliche Mineralisationsraten und Umsetzungszeiten aufweisen, kann eine langfristige Stabilisierung von OS erst nach vielen Jahren oder Jahrzehnten erreicht werden (*Janzen*, 2004). Deshalb ist die Einrichtung von Langzeitfeldexperimenten (LFE) mit konstanter Bewirtschaftung gefordert, um repräsentative Ergebnisse bezüglich OS und Nährstoffdynamik zu erhalten und die Bewirtschaftungsform im Hinblick auf ihre Auswirkungen auf Klima und Boden zu bewerten (*Watts* et al., 2010; *Zibilske* et al., 2002).

Der Fokus der vorliegenden Dissertation liegt auf dem Einfluss der Bodenbearbeitungsintensität auf Fraktionen der OS und auf Phosphorfraktionen, die unterschiedlich stabil im Boden vorliegen. Bisher sind nur wenige Informationen zum Einfluss der Bodenbearbeitungsintensität auf die C-, N- und P-Dynamik im Profil von Lössböden verfügbar, die durch ihre spezifischen physikalischen und chemischen Eigenschaften (*Dénes* et al., 1995) zu den fruchtbarsten Böden der Welt zählen (*Vitharana* et al., 2008). Die Ergebnisse der ersten beiden Teilexperimente dieser Dissertation stammen aus Untersuchungen von vier LFEs in Deutschland. Diese wurden zwischen 1990 und 1997 vom Institut für Zuckerrübenforschung (IfZ) in Zusammenarbeit mit der landwirtschaftlichen Abteilung der Südzucker AG eingerichtet. Die Versuchsflächen sind Lössstandorte

Zusammenfassung

(Phaeozeme und Luvisole) und liegen in Ost- und Süddeutschland (Friemar, Grombach, Lüttewitz, Zschortau). Alle vier Standorte umfassen eine einheitliche Fruchtfolge (Winterweizen – Winterweizen – Zuckerrübe) sowie drei Bodenbearbeitungsvarianten. Im Direktsaatverfahren (NT) wird keine Bodenbearbeitung durchgeführt, mit Ausnahme einer Saatbettbereitung mit einer Bearbeitungstiefe von 5 cm bevor Zuckerrüben gesät werden. Als reduzierte Bearbeitung (RT) wird ein Mulchsaatverfahren durchgeführt, bei dem mit einer Kreiselegge die obersten 15 cm des Bodens bearbeitet werden. Im konventionellen Pflugverfahren (CT) wird der Boden mit einem Scharpflug bis zu einer Tiefe von 25 bis 30 cm bearbeitet. Die Abbaudynamik von OS, speziell von verschiedenen Pflanzenresten, wurde basierend auf einem Inkubationsexperiment, in dem v.a. die Verteilung der Pflanzenreste im Boden im Fokus stand, untersucht.

Bewirtschaftungsformen wie z.B. Bodenbearbeitungsintensität sind bekannt dafür, die Umsatzdynamik der gesamten OS Substanz im Boden zu beeinflussen. Allerdings sind nur wenige Informationen zum Einfluss der Bearbeitungsintensität auf die einzelnen Fraktionen der OS mit unterschiedlichen Umsatzzeiten in Ober- und Unterböden verfügbar. Ein Teilexperiment dieser Dissertation untersucht die Auswirkung vom Direktsaatverfahren (NT), von reduzierter Bearbeitung (RT) und von konventioneller Bearbeitung (CT) auf den labilen (durchschnittliche Verweildauer <10 Jahre), intermediären und passiven (durchschnittliche Verweildauer >100 Jahre) Kohlenstoff- (C) und Stickstoff- (N) Pool sowohl in Ober- als auch in Unterböden. Dafür wurden Bodenproben vom Oberboden (0-5 cm) und Unterboden (5-25 cm) der drei Bearbeitungsvarianten (NT, RT, CT) von allen vier LFEs in Deutschland genommen. Die Größen der labilen, intermediären und passiven C- und N-Pools wurden durch ein Inkubationsexperiment (341 Tage) und einer physikalisch-chemischen Auftrennung bestimmt. In den Oberböden wurden höhere Vorräte der labilen C- und N-Pools für das NT-Verfahren (C: 1.76 t ha^{-1}, N: 166 kg ha^{-1}) im Gegensatz zum CT-Verfahren (C: 0.44 t ha^{-1}, N: 52 kg ha^{-1}) ermittelt.

Zusammenfassung

Im Gegensatz dazu waren die labilen C-Vorräte im Unterboden deutlich größer unter CT-Bearbeitung (C: 2.68 t ha^{-1}), verglichen zu NT- (2 t ha^{-1}) und RT-Verfahren (1.87 t ha^{-1}). Die Vorräte der intermediären C- und N-Pools waren um ein Vielfaches größer als die Vorräte der labilen und passiven C- und N-Pools. Im Ober- und Unterboden betrug der Anteil des intermediären C-Pool 73 bis 85% vom Gesamtvorrat des organischen C, für den intermediären N-Pool wurden 70 bis 95% des gesamten N-Vorrats berechnet. Nur für den Oberboden wurde ein Effekt der Bearbeitungsintensität auf den intermediären N-Pool festgestellt, der deutlich größer im NT- als CT-Verfahren war. Die Vorräte des passiven C- und N-Pools wurden zwar nicht durch die Bearbeitungsintensität beeinflusst, dafür wurde ein positiver Zusammenhang mit den Vorräten der gemessenen Tonfraktion und des bestimmten oxalat-löslichen Aluminiums festgestellt. Dies deutet auf einen starken Einfluss der standortspezifischen mineralischen Bodeneigenschaften auf die Größen der passiven C- und N-Pools. Insgesamt führt nach 14 bis 22 Jahren lediglich das Direktsaatverfahren verglichen mit dem konventionellen Pflugverfahren zu höheren Vorräten der labilen C- und N-Pools im Oberboden. Der Anstieg der labilen und intermediären Pools im Oberboden durch Reduzierung der Bodenbearbeitungsintensität sollte sich positiv auf den Nährstoffkreislauf und die Bodenproduktivität auswirken. Diese Erkenntnis ist von besonderer Bedeutung für die Entwicklung eines nachhaltig bewirtschafteten landwirtschaftlichen Ökosystems. Allerdings wurde keine Erhöhung des C- und N-Pools unter Direktsaatverfahren im Unterboden festgestellt. Offenbar ist der Einfluss des Bearbeitungssystems auf die OS-Dynamik unterschiedlich für den Ober- und Unterboden. Durch die tiefere Einarbeitung der Pflanzenreste waren die Vorräte des labilen C- und N-Pools im Unterboden unter CT größer, als unter RT und NT. Der intermediäre C- und N-Pool hat durch die hohen Anteile am gesamten C- und N-Pool eine deutlich größere Bedeutung für die OS-Dynamik als der labile und passive C- und N-Pool. Generell weisen die Ergebnisse dieses Teilexperiments auf eine Bodentiefe-

Zusammenfassung

abhängige Ausprägung der Reaktion der OS-Pools auf die Bodenbearbeitungsintensität hin. Daraus ergibt sich, dass die potentiellen Vorteile mit abnehmender Bodenbearbeitungsintensität bezüglich der Bodenfunktionen, die eng mit der Dynamik der OS verbunden sind, einzeln für Ober- und den Unterböden bewertet werden müssen. Zwar war der intermediäre N-Pool in diesem ersten Teilexperiment im Oberboden unter reduzierter Bodenbearbeitung am größten, allerdings scheint die Ausprägung dieses Effekts durch Bodenbearbeitung stark abhängig von den standortspezifischen Eigenschaften zu sein. Es könnte sein, dass es länger als 22 Jahre kontinuierlicher Bodenbearbeitung dauert, um die intermediären C- und N-Pools im Ober- und Unterboden entscheidend und standort-unabhängig zu beeinflussen. Der passive C- und N-Pool wurde weder im Ober- noch im Unterboden durch die Bodenbearbeitungsintensität beeinflusst, sondern scheint eher von mineralischen Bodeneigenschaften, wie z.B. dem Tongehalt beeinflusst zu sein.

Neben der OS-Dynamik kann auch Phosphor (P) im Boden durch das Bodenbearbeitungssystem beeinflusst werden. Als Hauptnährstoff ist P von Bedeutung für das Wachstum und die Gesundheit der Pflanzen. Durch eine unterschiedliche Bearbeitungsintensität und damit Bearbeitungstiefe werden Pflanzenreste ungleichmäßig tief im Boden verteilt. Außerdem wirkt sich das Bearbeitungssystem auf die Aggregatdynamik und den Bodenverlust durch Erosion aus, allerdings sind bisher nur wenige quantitative Daten erhoben. Ziel in diesem zweiten Teilexperiment war die Untersuchung des Effekts verschiedener Bodenbearbeitungssysteme auf die P-Verfügbarkeit. Die Ergebnisse basieren auf den Untersuchungen von Bodenproben der gleichen vier LFEs, die schon für das erste Teilexperiment verwendet wurden. Es wurde die Auswirkung des Direktsaatverfahrens (NT) mit der der konventionellen Bearbeitung (CT) auf die P-Verfügbarkeit verglichen. Phosphor im Boden wurde in einzelne Pools unterschiedlicher Stabilität unterteilt. Dafür wurde eine schrittweise Extraktionsmethode angewendet und

Zusammenfassung

in den Extrakten der Gehalt an anorganischem P (P_i) bestimmt. Zudem wurde der Gesamt-P-Gehalt (P_t) in den einzelnen Fraktionen durch den Aufschluss der entsprechenden Extrakte ermittelt, um aus der Differenz von P_t und P_i den Gehalt an organischem P (P_o) zu berechnen. Der P_t-Gehalt im Oberboden (0-5 cm) des NT-Verfahrens war mit 792 mg kg^{-1} etwa 15% höher verglichen zum CT-Verfahren und nimmt mit zunehmender Bodentiefe im NT-Verfahren stärker ab, als im CT-Verfahren. Dies galt auch für die anderen P-Fraktionen, ausgenommen der stabilsten P-Fraktion (Rest-P). Die höheren P-Gehalte im Oberboden des NT-Verfahrens resultieren wahrscheinlich aus der flacheren Einarbeitung von Pflanzenresten und P-Düngemitteln verglichen mit CT. Geschätzte Bodenverluste und in diesem Zusammenhang auch P-Verluste durch Wassererosion waren an allen vier Versuchsstandorten sehr gering. Es wurde ein positiver Zusammenhang zwischen den Gehalten an oxalat-löslichem Eisen sowie Aluminium und den labilen anorganischen P-Fraktionen festgestellt. Weiterhin wurde ein starker Zusammenhang zwischen den stabilen P-Fraktionen und den Tongehalten sowie den pH-Werten festgestellt. Generell hatte das Bodenbearbeitungsverfahren nur einen schwachen und nicht-signifikanten Effekt auf den Gesamt-P-Gehalt, wobei der P-Gehalt etwas höher in Böden des NT-Verfahrens als des CT-Verfahrens war. Dies resultierte aus einem Anstieg des labilen P-Gehaltes im Oberboden, vermutlich bedingt durch einen höheren Gehalt an organischem C in 0-5 cm. Korrelationsanalysen zeigten einen positiven Zusammenhang zwischen labilem P-Gehalt und dem Gehalt an organischem Kohlenstoff. Hingegen stand stabiles P in positiver Beziehung zu den Kalziumgehalten und sehr stabiles P (Rest-P) zu den Tongehalten. Durch die berechneten potentiell nur sehr geringen Erosionsverluste an den vier Standorten fehlen stärker ausgeprägte Effekte der Bearbeitungsintensität auf die P-Verfügbarkeit. Demnach wird der Erfolg eines Bodenbearbeitungssystems an den in diesem Experiment untersuchten

Zusammenfassung

Standorten mit den entsprechenden klimatischen Bedingungen und vorherrschenden Bodentypen von anderen Faktoren als der P-Verfügbarkeit bestimmt.

Neben der Bodenbearbeitungsintensität kann auch die Integration von Zwischenfrüchten in eine Fruchtfolge die C- und N-Dynamik, die Speicherung von C und N im Boden, sowie die Emission von Treibhausgasen beeinflussen. Entscheidende Faktoren können dabei die Art der Zwischenfrucht und ihre Biomasseproduktion, sowie die Einarbeitungs-tiefe der Zwischenfrüchte sein. Bisher wurden nur sehr wenige Daten zum Einfluss von Zwischenfrüchten auf die C- und N-Dynamik veröffentlicht. Ziel im dritten Teilexperiment dieser Dissertation war die Untersuchung von Bodenbearbeitungs- und Zwischenfruchteffekten auf labile C- und N-Pools während eines Inkubationsexperiments. Der Boden für dieses Experiment stammte von einer ackerbaulich genutzten Fläche in Nordhessen (Deutschland) und wurde in Acryzylinder (Säulen) gefüllt. In diese Säulen wurden drei Winterzwischenfrüchte gesät (Gelbsenf, Phacelia, Ölrettich), eine Kontrollvariante ohne Zwischenfrüchte wurde ebenfalls angelegt. Nach einer Wachstumsphase von 63 Tagen in Klimakammern mit simulierten Lichtverhältnissen und Temperaturverläufen (entsprechend Freilandbedingungen) wurden die Pflanzen abgeschnitten und in die obersten 15 cm eingearbeitet oder auf die Bodenoberfläche gegeben. Anschließend wurden die Säulen für 7 Tage bei -10 °C belassen, gefolgt von einer Inkubationsperiode von 80 Tagen bei 10 °C. Die Biomasseentwicklung war am größten in der Senf-Variante (7.0 t oberirdische Biomasse ha^{-1}), gefolgt von Phacelia (5.7 t ha^{-1}) und Ölrettich (4.4 t ha^{-1}). Die kumulativen CO_2- und N_2O-Emissionen unterschieden sich zwischen den Zwischenfruchtvarianten während der Inkubationszeit bei 10 °C nicht entscheidend. Auch wurden die CO_2- und N_2O-Emissionen nicht von der Einarbeitungstiefe der Pflanzenreste beeinflusst. Der Vorrat an kumulativ auswaschbarem mineralisiertem N (N_{min}) war am höchsten in den Kontrollsäulen ohne Zwischenfrucht. Die N_{min}-Vorräte in den Säulen mit Zwischenfruchtanbau waren in der Variante mit

Zusammenfassung

Einarbeitung der Pflanzenreste zwischen 51 und 72% niedriger als die entsprechende Kontrollvariante. In der Variante ohne Einarbeitung waren die N_{min}-Vorräte 36 bis 55% niedriger im Vergleich zur entsprechenden Kontrollvariante. Diese Ergebnisse zeigen den deutlich positiven Effekt des Zwischenfruchtanbaus in Hinsicht auf Vermeidung von Nitratverlusten durch Auswaschung zwischen Herbst und Frühling. Die Ergebnisse des dritten Teilexperiments der vorliegenden Dissertation deuten auf eine höhere C-Speicherung in Böden hin, deren Fruchtfolge den Anbau von Winterzwischenfrüchten beinhalten, verglichen mit Böden, in denen eine Bracheperiode enthalten ist. Dies resultiert aus einem höheren Gehalt an OS im Boden durch den Zwischenfruchtanbau, wodurch auch ein höherer organischer C-Gehalt resultiert. Es ist dabei zweitrangig, welche Zwischenfrucht angebaut wird. Insgesamt bestätigt dieses Experiment das Potential von Zwischenfrüchten zur temporären Speicherung von Nährstoffen verglichen mit brachen Böden, wobei weder die Zwischenfruchtart, noch ihre Einarbeitungstiefe die C- und N-Dynamik stark beeinflusst.

Insgesamt betrachtet zeigt die vorliegende Arbeit, das reduzierte Bodenbearbeitung zu einer C- und N-Speicherung im Boden führt. Eine Aussage, ob diese Speicherung durch Reduzierung der Bodenbearbeitungsintensität langfristig ist, konnte jedoch in der vorliegenden Dissertation nicht getroffen werden, da lediglich der Ist-Zustand nach 20-jähriger kontinuierlicher Bodenbearbeitung erhoben wurde. Weiterhin steht das Interesse an einer gesteigerten P-Verfügbarkeit im Gegensatz zu einer angestrebten höheren Konzentration an OS im Boden, da durch eine höhere Bearbeitungsintensität zwar P aus den stabileren Pools in die pflanzen-verfügbaren Pool transferiert wird, gleichzeitig aber auch der Gehalt an C und N in Form von klimarelevanten CO_2- und N_2O-Emissionen freigesetzt und N in Form von N_{min} ausgewaschen wird.

1 General introduction

Next to the interest in a higher quality of food, organic agriculture focus also on the stabilization and increase of organic matter (OM) and nutrient content in soil, as well as on soil productivity to maintain or increase the yields of arable crops. Furthermore, with the view on global warming, the sequestration of OM into soil may act opposite to the emissions of climate relevant greenhouse gases (*Franzluebbers*, 2005). The OM and nutrient content in arable soils is affected by management systems, such as tillage, fertilization and crop rotation (*Seiter & Horwath*, 2004). Because of various stability of OM and therefor different mineralization rates and turnover times, long-term stabilization of OM in soil will be reached only after a few years or decades (*Janzen*, 2004). Based on these findings, the initiation of long-term field trials with a constant management is required for reliable results concerning OM and nutrient dynamics to enable an evaluation of management strategies (*Watts* et al., 2010; *Zibilske* et al., 2002).

The key aspect in this dissertation is on the effect of tillage intensity on soil organic matter (SOM) pools as well as on soil phosphorus (P) fractions with different stability in soil. While the partition of SOM into pools with different turnover times has been under investigation either for fertilization (*Heitkamp* et al., 2009) or crop rotation treatments (*Kaiser* et al., 2007; *Rickman* et al., 2002), for (sandy) clay loam soils (*Bayer* et al., 2000; *Messiga* et al. 2011; *Zibilske & Bradford*, 2003) or loamy sand soils (*De Neve* et al., 2000; *Wright*, 2009), little data is available on the impact of tillage intensity on C, N, and P dynamic within the soil profile on loess soils. Soils derived from loess sediments are recognized as among the most fertile of the world (*Vitharana* et al., 2008), caused by its specific physical and chemical properties (*Dénes* et al., 1995).

This dissertation is based on results from a long-term tillage trial at four sites in Germany. The field experiment was initiated in 1990 by the Institute for Sugar Beet

Research (IfZ) in corporation with the Agricultural Division of the Südzucker AG. Detailed description is given in chapters 2.2.1 and 3.2.1. Results about the decomposition of OM based on different plant residues were provided by an incubation experiment, which is described in detail in chapter 4.2.

The following units of this general introduction mainly describe the physical, chemical and biological processes in soil and their affection by tillage intensity as well as the theoretical and methodological regard on measurement of SOM turnover and P availability in soil.

1.1 Impact of tillage intensity on OM transformation

The tillage induced turnover of SOM, either formation or mineralization of OM stocks, is affected by (i) litter placement in the soil, (ii) change of soil structure and (iii) impact on physical and climatic conditions in the soil (*Balesdent* et al., 2000; *Coppens* et al., 2006; *Vu* et al., 2009). Reduced tillage (RT) systems with a work depth of 15 cm, and no-till (NT) systems without tillage (direct drilling) results in decreased disruption of aggregates in soil (*Balesdent* et al., 2000). This is in contrast to conventional tillage (CT), including ploughing down to 25-30 cm soil depth. Several studies reports a lower OM stock in soils under CT caused by less physical protection and higher decomposition rates of OM compared to soils under RT and NT (e.g., *Mikha & Rice*, 2004; *Six* et al., 2000). Although, during the last years, a higher number of studies compared CT, RT and/or NT systems, information about the comparative impact of these three tillage systems on OM transformation in loess soils is scarce. Moreover, a comparison of different studies is complex, because country specific machinery is used for cultivation, which differs in tillage depth, operation occurrence and compacting of soil through machinery weight.

1   General introduction

1.2   Drivers for stabilization and destabilization of SOM and soil P

The stabilization of SOM increases with a lower decomposition rate, which depends mainly on (i) spatial inaccessibility and (ii) interactions between organic molecules and mineral constituents (*von Lützow* et al., 2006; *Kögel-Knabner* et al., 2008; *Schmidt* et al., 2011). Depending on tillage intensity and parent material these impact factors vary between surface and subsurface soil layers. Spatial inaccessibility results from the occlusion of SOM within aggregates and large organic macromolecule formation and thus its protection against microbial decomposition (*Six* et al., 2002). Organo-mineral interactions are polyvalent cation bridges, building of complexes between SOM and metal ions, e.g. aluminum (Al) and iron (Fe) ions (*von Lützow* et al., 2006). These different interactions affects the decomposition potential of SOM and leads to the formation of OM pools with different stability. *Von Lützow* et al. (2008) described a (i) labile (active) SOM pool with a turnover time with a maximum of 10 years, (ii) intermediate pool, with a turnover time of 10 up to 100 years and (iii) passive pool with a turnover time of more than 100 years.

Next to reduced tillage intensity, also the integration of cover crops into a crop rotation increases OM concentration in soil. The crop residues mainly belongs to the labile OM pool. Furthermore, cover crop cultivation tends to decrease potential soil erosion losses and has a positive effect on the nutrient balance in the soil (fixing of nutrients).

The main interest of organic agriculture to SOM is to increase the intermediate and passive OM pool to sequestrate C and N in soil. A tillage induced higher decomposition rate of OM to increase the concentration of plant available P in soil is in contrast to the main interest in the stabilization of OM, as P is an essential nutrient for all organisms. Because P do not exist in a gaseous state and is known to be very immobile in soil (*Deubel* et al., 2011; *Sinaj* et al., 2002), the shifting of P from more stable P pools into labile P pools is of main interest for plant nutrition. This labile P comprise P in the soil solution or P adhered on surfaces of more crystalline compounds (*Hedley* et al.,

1982; *Tiessen* & *Moir*, 1993; *Redel* et al., 2007), while P adsorbed to clay minerals and associated with mostly amorphous and some crystalline Fe and Al oxides (*Tiessen* et al., 1984) as well as P bound to calcium (Ca) ions (*Hedley* et al., 1982; *Guo* et al., 2000; *Frizano* et al., 2002) contributes to the stable P pool. The most stable P, namely residual-P is present in soil in the form of stable Al- and Fe-phosphates (*Hedley* et al., 1982; *Cross* & *Schlesinger*, 1995). The mobilization of P from the stable to the labile P pool can be reached by (i) the mineralization of P containing OM (*Ding* et al., 2011), (ii) change of ionic strength and cation composition of the electrolyte solution, which affects the P sorption, e.g. to clay, Fe, Al and Ca (*Tischner*, 1999), and (iii) higher concentrations of P solubilizing microorganisms (*Walpola* & *Yoon*, 2013), which results in P release. The tillage system affects the distribution of P in the soil profile through affecting the soil structure and SOM distribution. For study sites with a slope ≥ 5%, the tillage intensity has also a marked effect on the P dynamic and distribution through high soil erosion losses, because eroded soil material contains P adhered on clay minerals or surfaces crystalline compounds (*Schwertmann* et al., 1987).

1.3   SOM pools with different turnover times

Soil OM comprise organic molecules with different turnover times caused by various chemistry composition (*Helfrich*, 2006), which results in a different stabilization against microbial decomposition. The partition of OM into a small number of pools with different mean residence times (MRT) investigate a short or long-term stabilization of OM (*Janzen*, 2004).

Depending on the research question within an experiment, the OM have to be divided in the several pools by physical, chemical or kinetically fractionation. The interest in physical separated OM fractions mainly contributes to a possible mechanically impact of tillage systems on aggregate dynamics in soil (*Tisdall* & *Oades*, 1982). The physical division of OM into different density fractions was conducted in the present dissertation as a pre-treatment for a chemical fractionation following an approach of *Balesdent* et al. (1991) and *von Lützow* et al. (2007). They reported an overestimation of the most stable C and N pool (MRT > 100 years) without the removal of the light fraction (LF) of OM. The LF comprise the free and aggregate occluded OM, which consists of easily decomposable particles with a MRT of less than 10 years, and contributes to the labile OM. Following this approach, the C and N content determined in the heavy fraction (HF) which consists of OM associated with soil minerals (MRT > 100 years; *Helfrich* et al., 2007), was declared as the most stable C and N fraction.

A chemical fractionation via oxidation offer the determination of the passive C and N pool as a stable pool of C_{org}. *Helfrich* et al. (2007) developed a fractionation method, which was used in the sub-experiment described in chapter 2, including hydrogen peroxide (H_2O_2) and sodium peroxydisulfate ($Na_2S_2O_8$), buffered with sodium bicarbonate ($NaHCO_3$). The soil residue after the oxidation comprise SOM with associations of organic molecules with soil mineral compartments (*Kögel-Knabner* et al. 2008). It is therefor important to remove the $NaHCO_3$ buffer with hydrochloric acid (HCl, 0.01 M), to avoid an overestimation of the passive C pool by assign the carbonate C from the buffer to the passive C pool.

For the partition of SOM into the different pools, next to the chemical fraction, which enable the measurement of the passive C and N pools as described above, also the labile and intermediate OM pool has to be quantified. For this purpose, the implementation of mineralization experiments provide data, which can be used for mathematical description of OM pools, and can be induct to simple models of first-order kinetics, mostly one or two pool models. Because such mineralization experiments take about one year an the labile OM pool has a MRT of <10 years, the decomposition of only one part of the labile OM pool can be quantified via measurement of carbon dioxide

(CO_2) release and net N mineralization.

Each pool can be characterized by certain compounds, which decay by first order kinetics. The decay constant k varies between the OM pools with their stability and with environmental conditions amongst others temperature, moisture, and soil texture. Because for the very labile and labile OM pool the decay constant k is estimate with models implicated in different statistical programs, k vary slightly and therefore the estimation of the size of these two pools can be different. This is also true for the intermediate OM pool, which is calculated as the difference between total organic carbon (C_{org}) and the measured passive OM pool and the estimated labile (and very labile) OM pool. Only for the passive/stable OM pool, a number of separation methods are reported in literature. However, apart from methodological uncertainties, respectively disagreements, the usefulness of the partition of SOM into pools by mode-ling is a prerequisite to describe and predict SOM turnover (*von Lützow* et al., 2007).

1.4 Soil P pools with different stability

Phosphorus (P), as an essential nutrient for all organisms, exists in soil in many different organic and inorganic forms, but only a small part is plant available (*Redel* et al., 2007). As described above for SOM partition (chapter 1.3), also the fractionation of soil P into pools with different stability or plant availability supply information for a better understanding of P dynamic. Through increased tillage intensity, the reducing of SOM content (*D'Haene* et al., 2008) is accompany with reducing organically bound P, which results in higher P losses (*Scholz* et al., 2008). The P fractionation can be conduct by different methods, which leads to more than less similar P pools, but the extractants can differ. However, all methods differentiates between P pools according to their solubility in these extractants (*Hedley* et al. 1982; *O'Halloran*, 1993). Mostly, the P fractionation results in three main P pools, namely labile P, which is directly available for plants

1   General introduction

uptake (*Guo* et al., 2000, *Redel* et al., 2007), stable P may become plant available during short term soil transformations (*Guo* et al, 2000; *Frizano* et al., 2002), and the residual P is the most stable P fraction in soil, which is not plant available, possibly only after long-term soil transformations (*Hedley* et al., 1982; *Cross & Schlesinger*, 1994).

1.5 General Objectives

Overall, tillage systems and cover crops may affect the OM decomposition and distribution of P in different pools. To get an improved understanding on OM and P dynamics in soil affected by conventional, reduced and no-till systems, we used the information of four long-term tillage trials, and of an incubation experiment with different catch crops under controlled conditions. There is a lack of knowledge about the effect of different tillage systems on the OM dynamic in loess soils, especially the fractionation into pools with specific MRT in the surface and subsurface soil. Furthermore, for an improved understanding of P dynamics and control factors for P stabilization depending on tillage intensity in soil, information about tillage effect on different soil types is required. Thus, a P fractionation on loess soil is presented in this dissertation, because until now investigations on loess soils found less attention.

Specific objectives:

(i) Tillage impact of conventional, reduced and no-till systems on C and N mineralization in surface and subsurface soils, as well as relation of OM pools to soil texture and selected chemical soil properties (chapter 2);

(ii) Tillage impact of conventional and no-till systems on P dynamic, availability and stratification in the soil profile, as well as relation of P fractions to soil texture and selected chemical soil properties (chapter 3);

(iii) Impact of depth distribution of crop residues on the C and N mineralization of different winter cover crops (chapter 4).

1.6 References

Balesdent, J., Chenu, C., Balabane, M. (2000): Relationship of soil organic matter dynamics to physical protection and tillage. *Soil & Tillage Research* 53, 215–230.

Balesdent, J., Pétraud J. P., Feller C. (1991): Effets des ultrasons sur la distribution granulométrique des matières organiques des sols. *Science du Sol* 29, 95–106.

Bayer, C., Mielniczuk, J., Amado, T. J. C., Martin-Neto, L., Fernandes, S. V. (2000): Organic matter storage in a sandy clay loam Acrisol affected by tillage and cropping systems in southern Brazil. *Soil & Tillage Research* 54, 101–109.

Coppens, F., Merckx, R., Recous, S. (2006): Impact of crop residue location on carbon and nitrogen distribution in soil and in water-stable aggregates. *European Journal of Soil Science* 57, 570–582.

Cross, A. F., Schlesinger, W. H. (1995): A literature-review and evaluation of the Hedley fractionation - Applications to the biogeochemical cycle of soil-phosphorus in natural ecosystems. *Geoderma* 64, 197–214.

D'Haene, K., Vermang, J., Cornelis, W. M., Leroy, B. L. M., Schiettecatte, W., Neve, S. de, Gabriels, D., Hofman, G. (2008): Reduced tillage effects on physical properties of silt loam soils growing root crops. *Soil & Tillage Research* 99, 279–290.

De Neve, S., Hofman, G. (2000): Influence of soil compaction on carbon and nitrogen mineralization of soil organic matter and crop residues. *Biology and Fertility of Soils* 30, 544–549.

Deubel, A., Hofmann, B., Orzessek, D. (2011): Long-term effects of tillage on stratification and plant availability of phosphate and potassium in a loess chernozem. *Soil & Tillage Research* 117, 85–92.

Ding, X., Fu, L., Liu, C., Chen, F., Hoffland, E., Shen, J., Zhang, F., Feng, G. (2011): Positive feedback between acidification and organic phosphate mineralization in the rhizosphere of maize (Zea mays L.). *Plant and Soil* 349, 13–24.

Franzluebbers, A. J. (2005): Soil organic carbon sequestration and agricultural greenhouse gas emissions in the southeastern USA. *Soil & Tillage Research* 83, 120–147.

Frizano, J., Johnson, A. H., Vann, Scatena, F. N. (2002): Soil phosphorus fractionation during forest development on landslide scars in the Luquillo Mountains, Puerto Rico. *Biotropica* 34, 17–26.

Guo, F., Yost, R. S., Hue, N. V., Evensen, C. I., Silva, J. A. (2000): Changes in phosphorus fractions in soils under intensive plant growth. *Soil Science Society of America Journal* 64, 1681–1689.

Hedley, M. J., Stewart, J. W. B., Chauhan, B. S. (1982): Changes in inorganic and organic soil-phosphorus fractions induced by cultivation practices and by laboratory incubations. *Soil Science Society of America Journal* 46, 970–976.

Heitkamp, F., Raupp, J., Ludwig, B. (2009): Impact of fertilizer type and rate on carbon and nitrogen pools in a sandy Cambisol. *Plant and Soil* 319, 259–275.

Helfrich, M., Flessa, H., Mikutta, R., Dreves, A., Ludwig, B. (2007): Comparison of chemical fractionation methods for isolating stable soil organic carbon pools. *European Journal of Soil Science* 58, 1316–1329.

Janzen, H. H. (2004): Carbon cycling in earth systems - a soil science perspective. *Agriculture Ecosystems & Environment* 104, 399–417.

Kaiser, K., Mikutta, R., Guggenberger, G. (2007): Increased stability of organic matter sorbed to ferrihydrite and goethite on aging. *Soil Science Society of America Journal* 71, 711–719.

Kögel-Knabner, I., Guggenberger, G., Kleber, M., Kandeler, E., Kalbitz, K., Scheu, S., Eusterhues, K., Leinweber, P. (2008): Organo-mineral associations in temperate soils: Integrating biology, mineralogy, and organic matter chemistry. *Journal of Plant Nutrition and Soil Science* 171, 61–82.

Lóczy, D., Szalai, L. (1995): Assessment of loess as parent material for agroecological potential. *GeoJournal* 36, 275–280.

Messiga, A. J., Ziadi, N., Angers, D. A., Morel, C., Parent, L.-E. (2011): Tillage practices of a clay loam soil affect soil aggregation and associated C and P concentrations. *Geoderma* 164, 225–231.

Mikha, M. M., Rice, C. W. (2004): Tillage and manure effects on soil and aggregate-associated carbon and nitrogen. *Soil Science Society of America Journal* 68, 809–816.

O'Halloran, I. P. (1993): Effect of tillage and fertilization on inorganic and organic soil phosphorus. *Canadian Journal of Soil Science* 73, 359–369.

Redel, Y. D., Rubio, R., Rouanet, J. L., Borie, F. (2007): Phosphorus bioavailability affected by tillage and crop rotation on a Chilean volcanic derived Ultisol. *Geoderma* 139, 388–396.

Rickman, R., Douglas, C., Albrecht, S., Berc, J. (2002): Tillage, crop rotation, and organic amendment effect on changes in soil organic matter. *Environmental Pollutions* 116, 405–411.

Schmidt, M. W. I., Torn, M. S., Abiven, S., Dittmar, T., Guggenberger, G., Janssens, I. A., Kleber, M., Kögel-Knabner, I., Lehmann, J., Manning, D. A. C., Nannipieri, P., Rasse, D. P., Weiner, S., Trumbore, S. E. (2011): Persistence of soil organic matter as an ecosystem property. *Nature* 478, 49–56.

Scholz, G., Quinton, J. N., Strauss, P. (2008): Soil erosion from sugar beet in Central Europe in response to climate change induced seasonal precipitation variations. *Catena* 72, 91–105.

Schwertmann, U., Vogl W., Kainz, W. (1987): Bodenerosion durch Wasser. Vorhersage des Abtrags und Bewertung von Gegenmaßnahmen. Verlag Eugen Ulmer, Stuttgart (Germany).

Seiter, S., Horwath, W. R. (2004): Strategies for managing soil organic matter to supply plant nutrients. In: *Magdoff, F., Weil, R. R.*: Soil organic matter in sustainable agriculture. CRC Press LLC, Boca Raton, London, New York, Washington D.C., pp. 268–293.

Sinaj, S., Stamm, C., Toor, G. S., Condron, L. M., Hendry, T., Di, H. J., Cameron, K. C., Frossard, E. (2002): Phosphorus exchangeability and leaching losses from two grassland soils. *Journal of Environmental Quality* 31, 319–330.

Six, J., Conant, R. T., Paul, E. A., Paustian, K. (2002): Stabilization mechanisms of soil organic matter: Implications for C-saturation of soils. *Plant and Soil* 241, 155–176.

Six, J., Elliott, E. T., Paustian, K. (2000): Soil macroaggregate turnover and microaggregate formation: a mechanism for C sequestration under no-tillage agriculture. *Soil Biology & Biochemistry* 32, 2099–2103.

Tiessen, H., Moir, J. O. (1993): Characterization of available P by sequential extraction. In: Soil sampling and methods of analysis. Lewis Publishers, Boca Raton.

Tiessen, H., Stewart, J. W. B., Cole, C. V. (1984): Pathways of phosphorus transformations in soils of differing pedogenesis. *Soil Science Society of America Journal* 48, 853–858.

Tischner, T. (1999): Investigations of phosphorus leaching from sandy soil. In: *Heathwaite, A. L.*: Impact of land-use change on nutrient loads from diffuse sources. Proceedings of an international symposium held during IUGG 99, the XXII General Assembly of the International Union of Geodesy and Geophysics, at Birmingham, UK 18-30 July 1999. IAHS, Wallingford, pp. 51–56.

Tisdall, J. M., Oades, J. M. (1982): Organic-Matter and Water-Stable Aggregates in Soils. *Journal of Soil Science* 33, 141–163.

Vitharana, U. W. A., van Meirvenne, M., Simpson, D., Cockx, L., Baerdemaeker, J. de (2008): Key soil and topographic properties to delineate potential management classes for precision agriculture in the European loess area. *Geoderma* 143, 206–215.

von Lützow, M., Kögel-Knabner, I., Ekschmitt, K., Matzner, E., Guggenberger, G., Marschner, B., Flessa, H. (2006): Stabilization of organic matter in temperate soils: mechanisms and their relevance under different soil conditions - a review. *European Journal of Soil Science* 57, 426–445.

von Lützow, M., Kögel-Knabner, I., Ekschmitt, K., Flessa, H., Guggenberger, G., Matzner, E., Marschner, B. (2007): SOM fractionation methods: Relevance to functional pools and to stabilization mechanisms. *Soil Biology & Biochemistry* 39, 2183–2207.

von Lützow, M., Kögel-Knabner, I., Ludwig, B., Matzner, E., Flessa, H., Ekschmitt, K., Guggenberger, G., Marschner, B., Kalbitz, K. (2008): Stabilization mechanisms of organic matter in four temperate soils: Development and application of a conceptual model. *Journal of Plant Nutrition and Soil Science* 171, 111–124.

Walpola, B. C., Yoon, M.-H. (2013): Isolation and characterization of phosphate solubilizing bacteria and their co-inoculation efficiency on tomato plant growth and phosphorous uptake. *African Journal of Microbiology Research* 7, 266–275.

Watts, D. B., Torbert, H. A., Prior, S. A., Huluka, G. (2010): Long-Term Tillage and Poultry Litter Impacts Soil Carbon and Nitrogen Mineralization and Fertility. *Soil Science Society of America Journal* 74, 1239–1247.

Wright, A. L. (2009): Phosphorus sequestration in soil aggregates after long-term tillage and cropping. *Soil & Tillage Research* 103, 406–411.

Zibilske, L. M., Bradford, J. M. (2003): Tillage effects on phosphorus mineralization and microbial activity. *SOIL SCIENCE* 168, 677–685.

Zibilske, L. M., Bradford, J. M., Smart, J. R. (2002): Conservation tillage induced changes in organic carbon, total nitrogen and available phosphorus in a semi-arid alkaline subtropical soil. *Soil & Tillage Research* 66, 153–163.

2 Impact of tillage intensity on carbon and nitrogen pools in surface and subsurface soils of four long-term field experiments

Michael Kaiser[1*], Christiane Piegholdt[1], Rouven Andruschkewitsch[1], Heinz-Josef Koch[2], Bernard Ludwig[1]

[1]*Department of Environmental Chemistry, University of Kassel, Nordbahnhofstr. 1a, 37213 Witzenhausen, Germany*

[2]*Institute of Sugar Beet Research, Holtenser Landstr. 77, 37079 Göttingen, Germany*

*corresponding author: michael.kaiser@uni-kassel.de

Key words: long-term field experiment; conventional, reduced, and no tillage, mineralization experiment; labile, intermediate, and passive C and N pools

Abstract

Management options such as the intensity of tillage are known to influence the turnover dynamics of soil organic matter. However, less information is available about the influence of the tillage intensity on individual soil organic matter pools of different turnover dynamics in surface and subsurface soils. This study aimed to analyze the impact of no tillage (NT), reduced tillage (RT), and conventional tillage (CT) on labile (mean residence time <10 years), intermediate, and passive (mean residence time >100 years) C and N pools in surface soils compared to subsurface soils. We took surface (0-5 cm) and subsurface (5-25 cm) soil samples from NT, RT, and CT tillage systems of four long-term field experiments in Germany. The labile, intermediate and passive C and N pool sizes were determined based on the combined application of a decomposition experiment (~ 1 year) and a physical-chemical separation procedure. For the surface soils, we found higher stocks of the labile C and N pool under NT (C: 1.8 t ha^{-1},

2 Impact of tillage intensity on carbon and nitrogen pools in surface and subsurface soils of four long-term field experiments

N: 170 kg ha^{-1}) compared to CT (C: 0.4 t ha^{-1}, N: 60 kg ha^{-1}). In contrast, we detected significantly higher stocks of the labile C pool under CT (C: 2.7 t ha^{-1}) compared to NT (2 t ha^{-1}) and RT (1.9 t ha^{-1}) for the subsurface soils. Most of C and N were located in the intermediate pool and accounted for 73 to 85% of the C stocks and for 70% to 95% of the N stocks for the surface and subsurface soils. However, only for the surface soils, the stocks of the intermediate N pool were affected by the tillage intensity and distinctly larger for NT than for CT. The stocks of the passive C and N pool were not affected by the tillage intensity but were positively correlated to the stocks of the clay size fraction and oxalate soluble Al indicating a strong influence of site specific mineral characteristics on the size of these pools. Our results imply soil depth specific variations in the response of organic matter pools to tillage of different intensity. This indicates that the potential benefits of decreasing tillage intensity with respect to soil functions that are closely related to organic matter dynamics have to be evaluated separately for surface and subsurface soils.

2.1 Introduction

The global increase in concentrations of atmospheric greenhouse gases requires a reassessment of management practices (i.e., cropping, tillage, fertilization) to retain or accumulate organic carbon (C_{org}) and nitrogen in soils. The conversion of intensive to conservation tillage systems, for example, was shown to increase the C_{org} and total nitrogen (N_t) contents of surface soils (*Salinas-Garcia* et al, 1997; *Watts* et al., 2010) by lowering the decomposition rates of organic matter (OM) (*Kladivko*, 2001; *Zibilske* et al., 2002). In general, compared to intensive tillage, the lower physical impact of reduced or no tillage systems leads to less mechanical disruption of soil aggregates and, therefore, improved physical protection of OM against microbial decomposition (*Balesdent* et al.,

2 Impact of tillage intensity on carbon and nitrogen pools in surface and subsurface soils of four long-term field experiments

2000; *Cambardella* & *Elliott*, 1993; *Mikha* & *Rice*, 2004; *Six* et al., 2000; *Tisdall* & *Oades*, 1982; *Zotarelli* et al., 2007).

The tillage intensity also affects litter placement and thus decomposition dynamics (*Coppens* et al., 2006; *Hermle* et al., 2008; *Jacobs* et al., 2010; *Oorts* et al., 2007). Crop residues may accumulate on the soil surface of conservation tillage systems due to a reduced contact to the soil microbial community. Furthermore, changes in the micro-climatic conditions due to drying and rewetting are much more severe on the soil surface than deeper in the soil. It is well known that drying as well as rewetting induce stress on microorganisms and their metabolism (*Fierer* & *Schimel*, 2002) decreasing the microbial activity and OM decomposition. In contrast, the litter distribution in ploughed soil layers is relatively uniform and OM decomposition rates can be higher compared to reduced tillage (*Brye* et al., 2006; *Curtin* et al., 2000). Overall, different tillage systems may affect surface and subsurface soils differently, mainly due to different locations of residue input and subsequent hotspots of decomposition. Thus, it is necessary to analyze their impact on OM dynamics for surface and subsurface soils separately. However, knowledge about the influence of, for example, no-tillage, reduced tillage and conventional tillage on OM decomposition and stabilization processes in surface soils compared subsurface soils is scarce.

The contents of N_t and C_{org} comprise OM pools of different mean residence times (MRT) resulting from differences in their stabilization against microbial decomposition. Based on their different turnover dynamics, such pools have different ecological functions. The labile OM pool (i.e., MRT <10 years) is highly important for the nutrient cycling and the productivity of agricultural ecosystems, whereas the stable or passive OM pool (i.e., MRT >100 years) is crucial for the long-term sequestration of organic C in agricultural soils and mitigating climate change (*Janzen*, 2004). The transitional, intermediate OM pool (i.e., MRT 10-100 years) is probably important for both soil

2 Impact of tillage intensity on carbon and nitrogen pools in surface and subsurface soils of four long-term field experiments

productivity and long-term C storage. Changes in OM pools of different ecological functions resulting from changes in tillage intensity will have specific consequences for the services of agricultural ecosystems. A management induced increase of the labile OM compartment, for example, may lead to increased CO_2 emissions and/or higher productivity. However, for now, there is less cohesive information about the influence of tillage intensity on the labile, intermediate and stable/passive soil OM pools in surface soils compared to subsurface soils.

The labile soil C pool representing easily bio-available organic compounds can be quantified, for example, within laboratory incubation experiments by measuring the CO_2 emissions and the N mineralization (N_{min}) rates over a certain amount of time (e.g., *Heitkamp* et al., 2009; *Kader* et al., 2010). A highly important sub compartment of the labile OM pool represents the microbial biomass because this parameter quantifies on the one hand the potentially active decomposer community responsible for the OM decomposition and on the other hand a readily available energy source (e.g., dead microorganisms, extra- and intracellular compounds) for microorganisms (*von Luetzow* et al., 2007).

Stabilization of soil OM (i.e., MRT >100 years) results in temperate aerated topsoils from interactions between organic molecules and mineral constituents (e.g., oxides, polyvalent cations, Al-Silicates) and the occlusion of OM in aggregates (*von Lützow* et al., 2006; *Kögel-Knabner* et al., 2008; *Schmidt* et al., 2011). To quantify the stabilized OM compartment (i.e., passive pool), the intermediate and labile compartments have to be removed from the sample. *Helfrich* et al. (2007) combined a density fractionation with a chemical oxidation treatment and showed for passive OM pools of soils under different land use mean residence times distinctly larger than 500 years. It is known that the stable OM pool is heavily affected by site conditions such as soil mineral characteristics and pH (*Mikutta* et al., 2006; *Kaiser* et al., 2012) but the effects of the

2 Impact of tillage intensity on carbon and nitrogen pools in surface and subsurface soils of four long-term field experiments

tillage intensity on the stabilized OM pool which, for example, largely affects the aggregate turnover, are still uncertain.

The intermediate OM pool with MRT's between 10 and 100 years can be regarded quantity-wise as the difference of the total soil OM, the labile OM pool, and the stable OM pool (*Heitkamp* et al., 2009). *Hermle* et al. (2008) showed effects on the intermediate OM pool but only for different land use types (i.e., arable vs. grassland soils) for moist and cold-temperate conditions in Switzerland. Respective information for different tillage systems in dryer and warmer climatic conditions prevailing in large agricultural areas worldwide is missing.

In this study, we aimed to analyze the impact of tillage intensity on the labile, intermediate, and stable OM pools. We took samples from 0-5 cm and 5-25 cm depth from soils of four long-term field experiments on loess soils showing differences in soil texture as well as in concentrations of Fe and Al oxides. Each long-term field experiment included 3 tillage systems of conventional tillage (CT), reduced tillage (RT), and no-tillage (NT). We quantified the labile OM pool of these samples within a one year incubation experiment. The labile pool was further characterized by the N_{min} rates and the amounts of microbial biomass C (C_{mic}) and N (N_{mic}). Furthermore, we quantified the intermediate and stable OM pools as described above. We hypothesize:

(i) an increase in the labile and intermediate C and N pools for no-till compared to conventional and reduced tillage systems independent from site characteristics;

(ii) a different response of labile and intermediate C and N pools to different tillage intensities for surface soils compared to subsurface soils;

(iii) the passive C and N pools are not affected by the tillage intensity but by site specific differences in soil mineral characteristics.

2 Impact of tillage intensity on carbon and nitrogen pools in surface and subsurface soils of four long-term field experiments

2.2 Materials and Methods

2.2.1 Study sites

Samples were taken from soils of four different long-term field experiments (LFE) in Germany near to the cities of Friemar (Thuringia), Grombach (Baden-Wuerttemberg), Lüttewitz (Saxony), and Zschortau (Saxony). The LFE´s were established by the Institute of Sugar Beet Research (Göttingen, Germany) in cooperation with the agricultural division of Südzucker AG (Mannheim, Germany) between 1990/91 (Grombach) and 1997/98 (Zschortau) (Table 1). The annual mean air temperature of the study sites ranges from 8.0 to 9.3 °C and the annual precipitation from 512 to 776 mm (data were provided by Deutscher Wetterdienst) (Table 1). General soil properties a given in Table 1. The soil texture varies between the four loess sites with, for example, clay amounts (0-25 cm) ranging from 464 t ha^{-1} (Zschortau) to 902 t ha^{-1} (Friemar).

Table 1: Site designations, year of establishment, altitude, and climatic conditions of the long-term field experiments and mean values of the soil sand, silt, and clay, as well as oxalate soluble Al and Fe (Al$_{ox}$, Fe$_{ox}$) stocks (0- ~25 cm) and the pH values. The soil data shown are mean values of three pseudo replicates and the standard errors are given in parentheses.

	Site characteristics				Soil properties						
Site	Year established	Altitude (m)	Temperature (°C[a])	Precipitation (mm[a])	Sand	Silt	Clay	Al$_{ox}$	Fe$_{ox}$	Soil type[b]	pH
							(t ha^{-1})				
Friemar (Thuringia)	1992/1993	310	8.0	554	139 (24)	1959 (76)	902 (52)	2.51 (0.06)	4.12 (0.49)	Haplic Phaeozem	7.13 (0.06)
Grombach (Baden-Wuerttemberg)	1990/1991	270	9.3	776	60 (9)	2167 (77)	774 (85)	2.04 (0.12)	4.95 (0.66)	Luvisol/ Phaeozem	6.34 (0.14)
Lüttewitz (Saxony)	1992/1993	290	8.6	572	53 (11)	2338 (3)	609 (10)	1.61 (0.08)	6.27 (0.12)	Luvisol	6.72 (0.10)
Zschortau (Saxony)	1997/1998	110	8.8	512	838 (30)	1699 (10)	464 (22)	1.60 (0.21)	4.67 (0.21)	Haplic Luvisol /Haplic Planosol	7.07 (0.05)

[a] long-term annual means as provided by Deutscher Wetterdienst
[b] according to the World Reference Base for Soil Resources (2006)

2 Impact of tillage intensity on carbon and nitrogen pools in surface and subsurface soils of four long-term field experiments

At all sites, the crop rotation consists of two growing seasons of winter wheat (*Triticum aestivum* L.), followed by sugar beet (*Beta vulgaris* ssp. *vulgaris* var. *altissima* DÖLL). Crop residues are left in the field. At Friemar and Grombach, winter wheat was sown in fall 2010 after previous wheat and harvested before soil sampling in fall 2011, while at Lüttewitz and Zschortau, sugar beets were sown in spring 2011, after a seedbed preparation down to 5 cm. At soil sampling in fall 2011, sugar beets were still in the field.

Each LFE consisted of one field, which was divided into three stripes of different tillage intensity ranging from 2.5 to 8 ha, respectively: i) conventional tillage (CT) managed with a moldboard plough down to 25-30 cm depth, ii) reduced tillage (RT) managed with a rigid-tine cultivator down to 15 cm depth, and iii) no-tillage (NT) without tillage, except for seedbed preparation with a rigid-tine cultivator or disc harrow to a depth of 5 cm before the sugar beets are sown. For the present study, we took soil samples from three pseudo field replicates of the CT, RT and NT treatments of each of the four LFE's. In each case composite soil samples consisting of five cores (core sampler, 4 cm diameter) were taken from 0-5 cm and 5-25 cm depth. Soil samples were taken in September 2011, sieved (<5 mm), and stored field moist at 4 °C.

2.2.2 Soil analyses

Field moist soil samples were analyzed for pH by extraction with $CaCl_2$ (20 g soil/50 ml 0.01 M $CaCl_2$). Dry samples were used to determine texture using the pipet method (*DIN ISO 11277*, 2002). Gravimetric soil moisture content was determined by drying samples at 105 °C for 24 h. Bulk density was determined according to *DIN ISO 11272* (1998). For the determination of oxalate extractable iron (Fe_{ox}) and aluminum (Al_{ox}), following *DIN 19684-6* (1997), a 5 g sample was shaken for 2 hours with 50 ml extraction solution (0.1 M ammonium oxalate and 0.1 M oxalic acid). After filtration through a fiberglass filter, the Al_{ox} and Fe_{ox} concentrations in the filtrates were measured by using an atomic

2   Impact of tillage intensity on carbon and nitrogen pools in surface and subsurface soils of four long-term field experiments

absorption spectrometer (for Al$_{ox}$ Model GBC 906 AAS, GBC Scientific Equipment, Braeside, Australia; for Fe$_{ox}$ Unicam 939 AAS with a Gilson 222 Rack 22 autosampler, Villiers, France).

Total C and N content of dry soil was determined by dry combustion (Elementar Vario El, Heraeus, Hanau, Germany). Carbonate-C (CO$_3$-C) in soil was determined to calculate the C$_{org}$ content as the difference between total C and CO$_3$-C. For the CO$_3$-C determination, following *DIN 19682-13* (2009), we used a Scheibler equipment, 5-10 g sample, and 20 ml of 10% HCl.

Contents of C$_{mic}$ and N$_{mic}$ were determined before the incubation experiment by chloroform fumigation extraction (*Vance* et al., 1987). Briefly, two portions of soil (5 g) were taken from each soil sample. One portion was directly extracted with 20 ml of 0.5 M K$_2$SO$_4$, the other subsample was extracted after fumigation with CHCl$_3$ for 24 h at 25 °C. After filtration of the suspensions (Whatman No. 595 ½), the extracts were frozen until measurement of C and N with a C/N analyzer (analytikjena multi N/C 2100S, Jena, Germany). Microbial biomass C and N was calculated as the difference between fumigated and unfumigated samples with conversion factors of 0.45 for C$_{mic}$ (*Joergensen*, 1996) and 0.54 for N$_{mic}$ (*Brookes* et al., 1985).

2.2.3 Incubation experiment

Net C and N mineralization was determined following the method developed by *Stanford & Smith* (1972). Briefly, duplicates of 200 g dry matter equivalent fresh soil sample (sieved <5 mm) were filled into plastic bottles with a volume of 250 ml. To get representative samples, we mixed the samples from 0-5 cm (and also from 5-25 cm) of the three pseudo field replicates of each tillage treatment and study site. The soil samples were brought to 60% of water holding capacity (WHC) with deionized (DI) water, then covered with a net to allocate the irrigation and placed in jars, which were connected over flexible tubes with a gas chromatograph. The samples were incubated in a climate chamber at 10 °C, which was about the annual mean temperature of the study sites.

After a pre-incubation for one week at 10 °C, every 4.5 h a gas sample was automatically taken with a P64 (Loftfields Analytische Lösungen, Neu Eichenberg, Germany) and analyzed for CO_2 with a gas chromatograph (Shimadzu Gas Chromatograph GC-14A, Duisburg, Germany; flow 2 ml min^{-1}). To determine the N_{min} production (i.e., NO_3^- and NH_4^+), the soil was irrigated with 400 ml 0.01 M $CaCl_2$ at first to remove all mineral N before the incubation started to make sure that only N will be measured, which was mineralized during the incubation period. A vacuum was applied to the bottles with flexible tubes and a pump to suck off the leachates and collect them in polyethylene bottles. The leachates were frozen until measurement of NO_3^- and NH_4^+ with a continuous flow analyzer (Evolution II auto-analyzer, Alliance Instruments, Cergy-Pontoise, France). Subsequent to the sampling of the leachate, we added of 25 ml N-free nutrient solution to avoid a suppression of the microbial activity by the limitation of nutrients and to recover the 60% of WHC. Leachates were sampled at first in a two-week interval and after 2 months in a 6-week interval. The greater intervals were chosen to provide a sufficient NO_3^- and NH_4^+ concentration for measurement in the leachates. During the decomposition experiment, we irrigated each soil sample 12 times. We finished the mineralization experiment after 341 days because the cumulative mineralization of C and N was well described by the applied one-pool model with $R^2 > 0.99$ ensuring the correct estimation of the decay constant k.

2.2.4 Physical-chemical fractionation

At first, the free and aggregate occluded organic particles were removed from the soil samples following an approach of *Balesdent* et al. (1991). This was done to avoid the mixing of C and N derived from labile pools with C and N of the passive OM pool

(*Jagadamma* et al., 2010). Free as well as aggregate occluded organic particles contribute to the easily decomposable amount of OM in topsoils (*Kaiser* et al., 2010). For the present study, we used a sodium polytungstate (SPT) solution (Sometu, Berlin, Germany) with a density of 1.8 g cm^{-3}. We added 40 ml of SPT solution to 10 g field moist soil (<5 mm) and applied 10 glass beads (5 mm diameter) to the suspension, which was shaken for 18 h at 175 rpm on a reciprocal shaker. Then, the suspension was centrifuged for 30 min. at 2000 x g and the supernatant was filtered through a polyamide filter (0.45 µm). The material on the filter (light fraction: <1.8 g cm^{-3}; LF) was washed with 2 l of DI water. The glass beads were removed and the soil pellet (heavy fraction: >1.8 g cm^{-3}; HF) was also filtered and washed with 2 l of DI water. The light and heavy fractions were dried for 48 h at 40 °C and the C and N concentrations of the HF fraction were determined via dry combustion.

After the separation of organic particles, we mixed 0.5 g of the HF with 250 ml of DI water and added 20 g of $Na_2S_2O_8$. The suspension was buffered with 22 g of $NaHCO_3$ and heated to 80 °C in a water bath with shaker function for 48 h. To provide a constant homogeneous sample distribution facilitating optimal oxidation conditions, we applied 80 glass beads (5 mm diameter) to the suspension. After oxidation, the glass beads were removed and each sample was washed two times with 40 ml of DI water, once with 40 ml of 0.01 M hydrochloric acid (HCl) to remove remaining carbonates from the $NaHCO_3$ buffer, and again twice with 40 ml DI water until a neutral pH was reached. After each washing, the suspension was centrifuged at 4000 x g for 20 min. and the supernatant was decanted. The cleaned extraction residue was dried at 40 °C and analyzed for C and N (passive C and N) concentration by dry combustion as described above.

2 Impact of tillage intensity on carbon and nitrogen pools in surface and subsurface soils of four long-term field experiments

2.2.5 Equivalent soil mass approach

The stocks of bulk soil OM and N_t as well as the labile and passive soil C and N pools, microbial eiomass C and N (C_{mic}, N_{mic}), mineralizable N (N_{min}), CO_2-C, the light and heavy fractions (LF, HF), clay and the oxalate soluble Fe and Al (Fe_{ox}, Al_{ox}) were calculated based on equivalent soil mass (*Ellert* et al., 2001). The stocks of individual parameters are given in t ha^{-1} or kg ha^{-1} for 500 t of surface soil (0 - ~5 cm) and for 2500 t of subsurface soil (5 - ~25 cm) using the measured depth specific concentration of the respective parameter (clay data are from mixed samples of 0-25 cm soil depth) and the depth specific bulk density. The stocks of intermediate C and N pools were calculated as the difference between the stocks of C_{org} and N_t and the stocks of labile and passive C and N, respectively.

2.2.6 Statistics and modeling

The data were analyzed with the GNU R Version 2.11.1 (*R Development Core Team*, 2010) by the Shapiro-Wilk normality test, analysis of variance (ANOVA) and correlation analysis. The data were analyzed as a split-plot design with tillage treatment as the main factor and soil depth as sub-factor. Because some data sets were not normally distributed, we conducted a logarithmic data transformation (boxcox transformation) to provide the preconditions (normal distribution and homogeneity of variance of the data sets) for a two-way ANOVA. Analysis of variance was performed on the averaged values of two subsamples. The four sites served as field replicates. Effects were considered to be significant at $p \leq 0.05$ and to be a trend at $p \leq 0.1$.

For correlation analyses, we used Spearman rank correlation to detect relationships between the amounts of the different C and N pools and the amounts of density fractions, microbial biomass, mineralized C and N as well as soil mineral characteristics. The correlations analyses were done for the combined data of surface and subsurface soils. For this we divided the soil mass equivalent amount of a parameter by the soil mass specific depth to facilitate a comparison on an amount per cm basis.

Modeling of C and N mineralization to estimate the size of the labile C and N pool was conducted with a one-pool model using the GNU R Version 2.11.1 (*R Development Core Team*, 2010). For the estimation of the decay constants we used a non-linear least square model with first-order compartment:

$$Y_{min}(t) = Y_l \times (1-\exp^{-k \times t}) \qquad (1)$$

where $Y_{min}(t)$ is C or N mineralized (kg ha^{-1}) (i.e., CO_2 or N_{min}) at time t (days), Y_l is the labile C or N pool (kg ha^{-1}), k is the decay constant (day^{-1}). To provide an unequivocal measure of soil C and N mineralization capacity, we followed the recommendation of *Wang* et al. (2003) and fitted Eq. (1) to the obtained data set of all tillage treatments and soil depths and fixed the decay constants as the average of the single decay constants of tillage treatments and soil depths (n = 6) to the obtained values. The obtained decay constants were: k = 0.0028 d^{-1} for C mineralization and k = 0.0011 d^{-1} for net N mineralization.

2.3 Results

2.3.1 Stocks of C_{org} and N_t

After 14 (Zschortau), 19 (Friemar, Lüttewitz), and 21 (Grombach) years of different tillage treatments, the stocks of C_{org} and N_t of the surface soils (0 - ~5 cm, based on an equivalent soil mass approach) were smaller for CT than RT and NT for all four sites (Table 2). Overall, the C_{org} and N_t stocks showed significantly higher stocks under NT (9 t C_{org} ha^{-1}, 0.8 t N_t ha^{-1}) than under CT (5.7 t C_{org} ha^{-1}, 0.6 t N_t ha^{-1}) in the surface soils, whereas stocks under RT did not differ significantly from the other tillage treatments (Table 3). The C_{org} and N_t stocks in 5 - ~25 cm as well as in 0 - ~25 cm depth showed no significant differences between the tillage systems.

Table 2: Site, tillage system, soil depth, and the respective stocks of organic C (C_{org}) and total N (N_t) as well as of the labile, intermediate, and passive C and N pools, the microbial biomass C and N (C_{mic}, N_{mic}), and the mineralized C (CO_2-C) and N (N_{min}). The data shown for passive C and N, C_{mic}, and N_{mic} are mean values of three pseudo replicates. The data shown for labile C and N, intermediate C and N, CO_2-C, and N_{min} are mean values of the two lab replicates (three pseudo replicates per site, treatment and depth were mixed for each lab replicate). Values in parentheses are standard errors.

[a] CT: conventional tillage
[b] RT: reduced tillage
[c] NT: no-till

Site	Tillage system	Soil depth	C_{org}	Labile C	Intermediate C	Passive C	C_{mic}	CO_2-C	N_t	Labile N	Intermediate N	Passive N	N_{mic}	N_{min}
		(cm)	(t ha-1)	(t ha-1)	(t ha-1)	(t ha-1)	(kg ha-1)	(t ha-1)	(t ha-1)	(kg ha-1)	(kg ha-1)	(kg ha-1)	(kg ha-1)	(kg ha-1)
Friemar	CT[a]	0-4.4	7.3 (0.3)	0.5 (0.1)	6.1 (0.3)	0.76 (0.04)	160 (20)	0.3 (0.1)	0.7 (0)	70 (0)	520 (10)	90 (10)	40 (10)	20 (0)
		5-24.0	35.6 (1.7)	2.9 (0.2)	29.6 (0.9)	3.09 (0.58)	750 (100)	1.8 (0.3)	3.3 (0.1)	350 (0)	2660 (10)	280 (10)	120 (30)	110 (0)
	RT[b]	0-5.0	10.2 (0.5)	1.1 (0.1)	8.5 (0.3)	0.53 (0.09)	200 (30)	0.7 (0.1)	0.9 (0)	110 (0)	750 (0)	50 (0)	40 (10)	30 (0)
		5-23.9	36.6 (1.5)	2.2 (0.2)	31.3 (1.2)	3.21 (0.12)	680 (70)	1.3 (0.3)	3.3 (0.1)	290 (40)	2660 (10)	340 (30)	130 (40)	90 (10)
	NT[c]	0-4.4	10.1 (0.3)	2.2 (0.1)	7.4 (0.1)	0.63 (0.09)	290 (0)	1.3 (0.1)	0.9 (0)	190 (10)	690 (0)	60 (10)	70 (10)	60 (0)
		5-23.0	33.5 (0.3)	2.0 (0.2)	27.5 (0.1)	4.08 (0.03)	680 (100)	1.2 (0.2)	3.1 (0.2)	240 (10)	2610 (10)	260 (0)	160 (60)	70 (0)
Grombach	CT	0-4.2	4.3 (0.2)	0.4 (0.1)	3.4 (0.1)	0.51 (0.05)	110 (20)	0.3 (0.1)	0.5 (0)	50 (10)	360 (0)	40 (0)	30 (10)	10 (0)
		5-24.5	20.4 (0.4)	2.3 (0.5)	15.2 (0.6)	2.89 (0.43)	550 (100)	1.4 (0.7)	2.0 (0)	260 (20)	1510 (0)	260 (20)	150 (50)	80 (10)
	RT	0-4.9	7.9 (0.5)	1.3 (0.1)	6.0 (0.3)	0.62 (0.09)	300 (30)	0.8 (0.1)	0.7 (0)	130 (10)	560 (20)	50 (10)	70 (10)	40 (0)
		5-22.7	31.9 (3.8)	2.0 (0.2)	24.6 (3.5)	2.94 (0.16)	630 (80)	1.2 (0.2)	2.9 (0.1)	250 (20)	2400 (10)	260 (10)	160 (40)	70 (10)
	NT	0-4.4	10.2 (0.8)	1.6 (0.1)	7.5 (0.6)	0.61 (0.08)	300 (20)	1.0 (0.1)	0.9 (0)	130 (20)	710 (10)	50 (10)	70 (10)	40 (10)
		5-22.2	29.9 (5)	2.0 (0.3)	21.5 (4.4)	2.99 (0.32)	620 (90)	1.3 (0.3)	2.7 (0.1)	250 (0)	2130 (50)	300 (50)	150 (40)	70 (0)
Lüttewitz	CT	0-3.7	5.6 (0.2)	0.4 (0.05)	4.6 (0.1)	0.54 (0.07)	110 (40)	0.3 (0.1)	0.6 (0)	50 (10)	490 (10)	20 (0)	30 (10)	10 (0)
		5-25.0	28.3 (0.7)	2.9 (0.35)	22.9 (0)	2.44 (0.33)	560 (220)	1.8 (0.4)	2.9 (0.1)	390 (10)	2370 (10)	90 (20)	120 (90)	120 (0)
	RT	0-4.6	8.3 (0.3)	1.6 (0.11)	6.2 (0.1)	0.5 (0.04)	350 (20)	1.0 (0.1)	0.8 (0)	210 (10)	540 (10)	20 (0)	70 (10)	60 (0)
		5-23.7	25.5 (0.8)	1.9 (0.56)	21.3 (0.1)	2.41 (0.28)	700 (20)	1.2 (0.7)	2.5 (0)	210 (10)	2230 (10)	90 (10)	180 (30)	60 (0)
	NT	0-4.1	8.6 (0.2)	1.5 (0.12)	6.6 (0)	0.5 (0.01)	300 (10)	0.9 (0.2)	0.8 (0)	190 (0)	620 (0)	20 (0)	70 (10)	60 (0)
		5-22.9	26.9 (0.3)	1.9 (0.17)	22.8 (0)	2.19 (0.14)	460 (10)	1.2 (0.2)	2.6 (0.1)	260 (20)	2260 (10)	100 (10)	150 (40)	80 (10)
Zschortau	CT	0-3.7	5.7 (0.4)	0.4 (0)	4.8 (0.3)	0.48 (0.06)	140 (10)	0.3 (0.1)	0.5 (0)	50 (0)	450 (0)	20 (0)	30 (10)	10 (0)
		5-24.5	29.0 (1.9)	2.5 (0.2)	24.0 (1.5)	2.48 (0.13)	750 (90)	1.6 (0.3)	2.6 (0.1)	280 (30)	2250 (40)	90 (10)	180 (50)	80 (10)
	RT	0-4.2	7.5 (0.1)	1.0 (0)	6.1 (0.1)	0.49 (0.02)	190 (20)	0.6 (0.1)	0.7 (0)	90 (10)	580 (20)	10 (0)	50 (0)	30 (0)
		5-24.6	30.3 (1.4)	1.5 (0.3)	26.7 (0.9)	2.21 (0.26)	520 (130)	0.9 (0.4)	2.8 (0.1)	110 (80)	2600 (80)	70 (10)	140 (50)	30 (20)
	NT	0-4.4	7.5 (0.1)	1.8 (0.1)	5.3 (0.1)	0.46 (0)	280 (70)	1.1 (0.1)	0.7 (0)	150 (30)	530 (30)	10 (0)	60 (20)	40 (10)
		5-23.4	25.7 (1.2)	2.1 (0.3)	21.3 (0.7)	2.27 (0.15)	480 (30)	1.3 (0.4)	2.4 (0.1)	220 (20)	2030 (40)	100 (20)	130 (20)	70 (10)

2   Impact of tillage intensity on carbon and nitrogen pools in surface and subsurface soils of four long-term field experiments

2.3.2 Microbial biomass C and N

At all four sites, contents of C_{mic} and N_{mic} in the surface soil were higher under NT and RT than under CT (Table 2). Overall, a trend of higher stocks of C_{mic} under NT (290 kg ha^{-1}) than under CT (130 kg ha^{-1}) was found for the surface soils and the N_{mic} stocks were significantly higher under NT (70 kg ha^{-1}) and RT (60 kg ha^{-1}) compared to CT (30 kg ha^{-1}) (Table 3). In the subsurface soil (5 - ~25 cm), the C_{mic} stocks ranged from 560 to 680 kg ha^{-1} and the N_{mic} stocks from 140 to 150 kg ha^{-1}, and there were no significant differences between the treatments (Table 3).

The LF stocks were between 4.9 t ha^{-1} (NT) and 2.3 t ha^{-1} (CT) in 0 - ~5 cm soil depth and between 10.9 t ha^{-1} (CT) and 9.3 t ha^{-1} (RT) in 5 - ~25 cm soil depth but we detected no significant effects of the tillage intensity neither for the surface soils nor for the subsurface soils. In contrast, we observed minor differences between the HF stocks of CT and NT for the soils in 0-5 cm depth (Table 3), whereas the HF stocks of soils in 5 - ~25 cm depth showed no effect of tillage intensity.

Table 3: Tillage system, soil depth, and the respective stocks of the soil organic C (C_{org}), total N (N_t), the light (LF) and heavy fraction (HF), the microbial biomass C and N (C_{mic}, N_{mic}). The data shown are mean values of the four study sites, the standard errors are given in parentheses. Values followed by different lower case letters are significantly different (p ≤0.05) and values followed by upper case letters are different by trend (p ≤0.1). Letters refer to the comparison of tillage treatments within one depth.

Tillage system	Soil depth	C_{org}	N_t	LF	HF	C_{mic}	N_{mic}
	(cm)	(t ha^{-1})	(t ha^{-1})	(t ha^{-1})	(t ha^{-1})	(kg ha^{-1})	(kg ha^{-1})
CT[a]	0- ~5	5.7 (0.6) b	0.6 (0.1) b	2.3 (0.2)	500 (0) a	130 (10) B	30 (0) b
	5- ~25	28.3 (3.1)	2.7 (0.3)	10.9 (1.8)	2490 (0)	660 (60)	140 (10)
RT[b]	0- ~5	8.5 (0.6) ab	0.8 (0.1) ab	4.2 (0.4)	500 (0) ab	260 (40) AB	60 (10) a
	5- ~25	30.5 (2.3)	2.9 (0.2)	9.3 (1.1)	2490 (0)	630 (40)	150 (10)
NT[c]	0- ~5	9.0 (0.6) a	0.8 (0.1) a	4.9 (0.9)	500 (0) b	290 (10) A	70 (10) a
	5- ~25	28.1 (1.8)	2.7 (0.2)	10.2 (1.2)	2490 (0)	560 (50)	150 (6)

[a]CT: conventional tillage, [b]RT: reduced tillage, [c]NT: no-till

2   Impact of tillage intensity on carbon and nitrogen pools in surface and subsurface soils of four long-term field experiments

2.3.3 CO_2 emission and net N mineralization

The amounts of mineralized C (CO_2-C) and N (N_{min}) of the surface soils after 341 days were markedly higher for NT and RT compared to CT (Table 2). In total, the amounts of cumulative emitted CO_2-C were significantly higher for NT (1.09 t ha^{-1}) and RT (0.78 t ha^{-1}) than for CT (0.28 t ha^{-1}) (Figure 1 a). In contrast, the CO_2-C emissions from the subsurface soils (5 - ~25 cm) were significantly higher for CT (1.66 t ha^{-1}) compared to NT (1.24 t ha^{-1}) and RT (1.16 t ha^{-1}) (Figure 1 c).

The N_{min} amounts were significantly higher for NT (51 kg ha^{-1}) compared to CT (16 kg ha^{-1}) for the soils of 0- ~5 cm depth (Figure 1 b). Similar to the cumulative CO_2-C emissions, we found the N_{min} amounts to be largest for CT (96 kg ha^{-1}) for the soils of 5-25 cm depth and we detected an effect of tillage intensity by trend between CT and RT (65 kg ha^{-1}) (Figure 1 d).

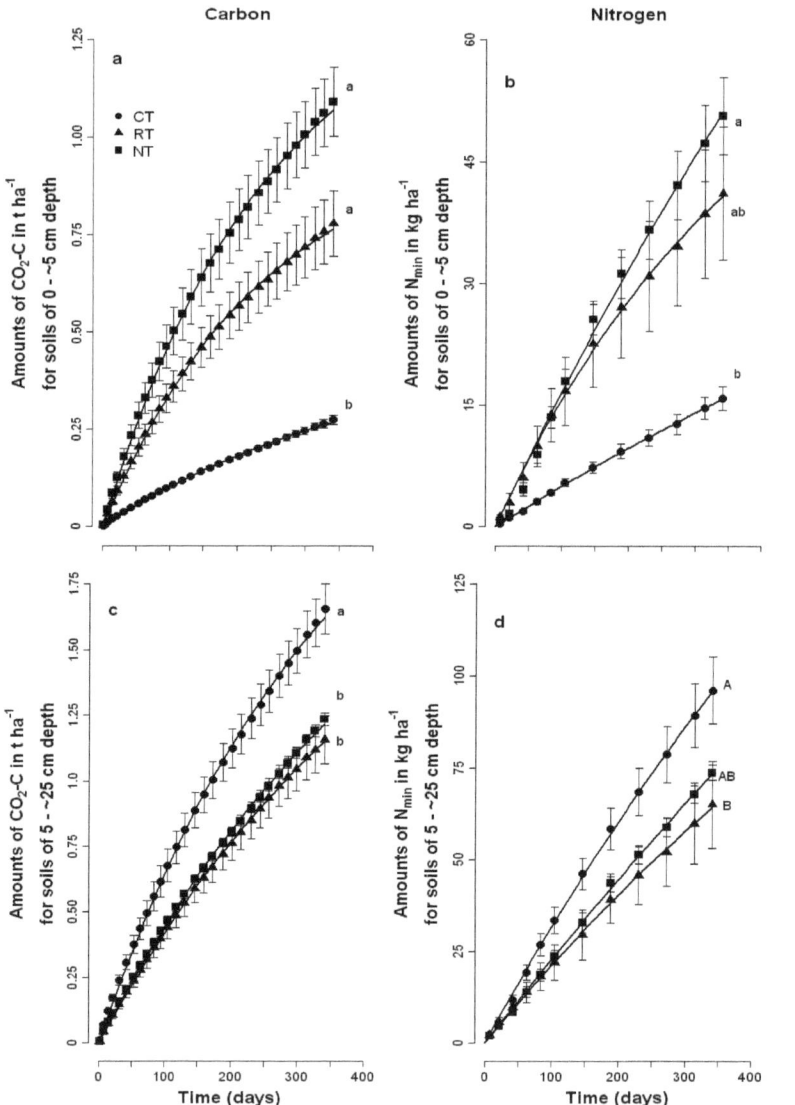

Figure 1: Cumulated stocks of emitted CO_2-C and mineralized net N (N_{min}) of soils in 0-~5 cm as well as 5- ~25 cm depth of the three tillage treatments (CT: conventional tillage, RT: reduced tillage, NT: no-till). Datapoints are means of the four study sites, error bars refer to standard errors of the means, and the solid lines refer to the interpolated data points with an exponential one-pool model. Values followed by different lower case letters are significantly different ($p \leq 0.05$) and values followed by upper case letters are different by trend ($p \leq 0.1$).

2   Impact of tillage intensity on carbon and nitrogen pools in surface and subsurface soils of four long-term field experiments

2.3.4   C and N pools

The cumulative CO_2-C emissions were well described by a simple one-pool model ($R^2 > 0.99$, $k = 0.0028$ d^{-1}). The modeled storage in the labile C pool ranged from 0.4 to 2.2 t ha^{-1} (7 to 24% of C_{org}) for the surface soil (0 - ~5 cm) and from 1.5 to 2.9 t ha^{-1} (5 to 11% of C_{org}) for the subsurface soil (Table 2). Overall, we detected significantly higher labile C stocks in the surface soils under NT (1.8 t ha^{-1}) and RT (1.3 t ha^{-1}) compared to CT (0.4 t ha^{-1}) (Figure 2 a). In contrast, the labile C pool was significantly higher under CT (2.7 t ha^{-1}) compared to NT (2 t ha^{-1}) and RT (1.9 t ha^{-1}) (Figure 2 c) in the subsurface soils.

The calculated stocks of the intermediate C pool ranged from 4.6 to 8.5 t ha^{-1} for the surface soils (73 to 84% of C_{org}) and from 15.2 to 31.3 t ha^{-1} for the subsurface soils (74 to 88% of C_{org}) (Table 2). The experimentally determined passive C pool stored between 0.5 and 0.8 t C ha^{-1} (5 to 12% of C_{org}) and between 2.2 and 4.1 t C ha^{-1} (9 to 14% of C_{org}) for the surface and subsurface soils, respectively (Table 2). Both the stocks of the intermediate and the passive C pools were not affected by the intensity of tillage.

The cumulative net N mineralization was well described by the applied one-pool model ($R^2 > 0.99$, $k = 0.0011$ d^{-1}). The modeled storage in the labile N pool ranged from 50 to 210 kg ha^{-1} for the surface soil (9-27% of N_t) and from 110 to 390 kg ha^{-1} for the subsurface soil (8-14% of N_t) (Table 2). Similar to the findings for labile C, the stocks of the labile N pool were significantly higher under NT (170 kg ha^{-1}) compared to CT (50 kg ha^{-1}) in 0-5 cm soil depth (Figure 2 b). In contrast and in agreement with the findings for labile C, the stocks of the labile N pool were largest under CT (320 kg ha^{-1}) in the subsurface soil and we found a trend of a higher stock under CT compared to RT (210 kg ha^{-1}) (Figure 2 d). The C/N ratios of the labile pool ranged from 7.4 to 13.8. The coefficients of the Spearman's rank correlation revealed significant relationships between the stocks of the labile C and N pools and the stocks of C_{mic} and N_{mic}, LF as well as CO_2-C and N_{min} (Figure 3 a-d and 4 a-d).

The stocks of the calculated intermediate N pool ranged from 360 to 750 kg ha^{-1} for the surface soils (70-88% of N_t) and from 1510 to 2660 kg ha^{-1} for the subsurface soils (75-95% of N_t) (Table 2). In contrast to the intermediate C pool, we found a trend of higher stocks of the intermediate N pool of NT (640 kg ha^{-1}) compared to CT (450 kg ha^{-1}) for the surface soils but no effect for the subsurface soils (Figure 2 b and d). The C/N ratios were similar to those of the labile pool and ranged from 9.4 to 11.7. We found significant relationships between the stocks of the intermediate C and N pools and the stocks of C_{mic}, LF, and N_{min} (Figure 3 e-g and 4 e-g).

The passive pool had a higher C/N ratio (ranging from 8.5 to 38.3) and stocks of the passive N pool ranged from 10 to 90 kg ha^{-1} and 70 to 340 kg ha^{-1} for the surface and subsurface soils, respectively, which accounts for 7-13% of N_t for each soil depth (Table 2). We detected no effects of the tillage intensity on the stocks of the passive N pool neither for the surface soils nor for the subsurface soils. The stocks of the passive C and N pools correlated with the stocks of Al_{ox} and clay (Figure 3 h and i, Figure 4 h and i).

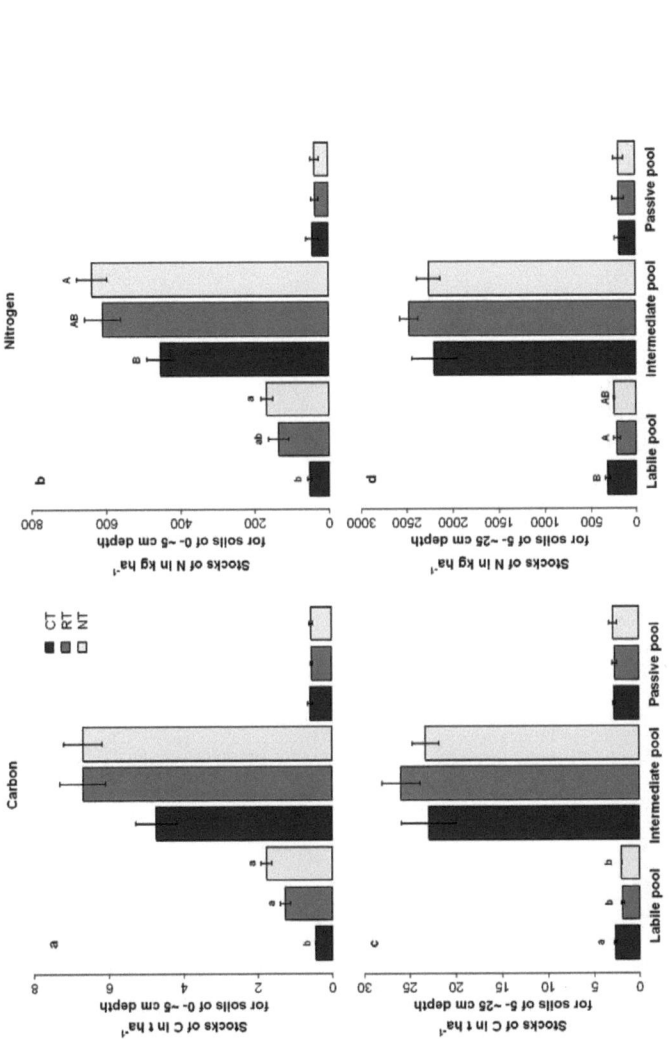

Figure 2: Stocks of the labile, intermediate, and passive C and N pools of soils in 0-~5 cm and 5-~25 cm depth of the three tillage treatments (CT: conventional tillage; RT: reduced tillage; NT: no-till). Columns show the means values of the four study sites, error bars refer to standard errors of the means. Values followed by different lower case letters are significantly different (p errors of the means. Values followed by different lower case letters are significantly different (p letters are different by trend (p

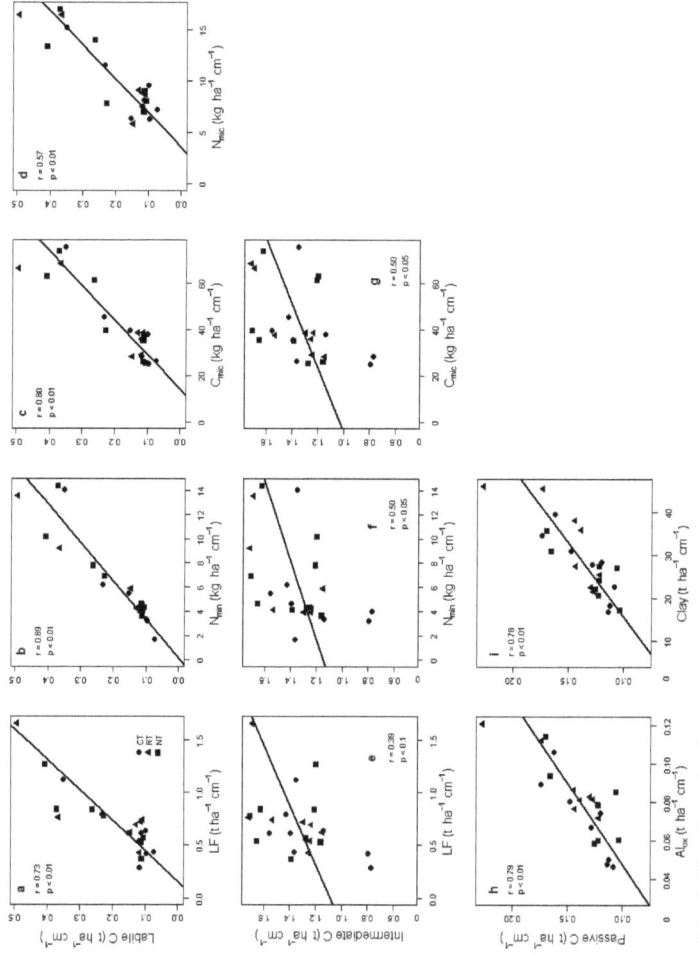

Figure 3: Stocks of the labile C pool (all soils) versus stocks of the (a) light fraction (LF), (b) mineralized N (N min), (c) microbial biomass C (Cmic), and (d) microbial biomass N (Nmic), stocks of the intermediate C pool (all soils) versus stocks of the (e) LF, (f) N min and (g) Cmic and stocks of the passive C pool (all soils) versus stocks of the (h) oxalate soluble Al (Al ox), and (i) clay fraction. Correlations were analyzed with Spearman's rank correlation test.

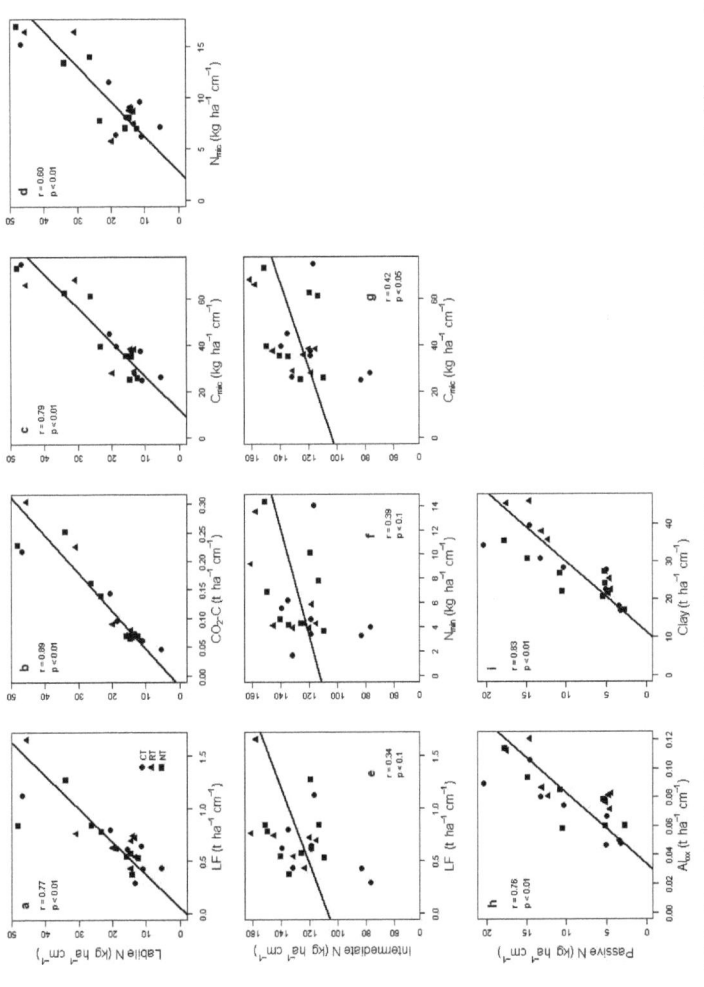

Figure 4: Stocks of the labile N pool (all soils) versus stocks of the (a) light fraction (LF), (b) CO_2-C, (c) microbial biomass C (C_{mic}), and (d) microbial biomass N (N_{mic}), stocks of the intermediate N pool (all soils) versus stocks of the (e) LF, (f) N_{min} and (g) C_{mic} and stocks the passive N pool (all soils) versus stocks of the (h) oxalate soluble Al (Al_{ox}), and (i) clay fraction. Correlations were analyzed with Spearman's rank correlation test.

2   Impact of tillage intensity on carbon and nitrogen pools in surface and subsurface soils of four long-term field experiments

2.4   Discussion

2.4.1   Stocks of C_{org} and N_t

Main input sources for C and N into the soil of the LFE's were fertilization (i.e., the same site specific N fertilization for NT, RT, and CT; no legumes were used in the crop rotations), which affects biomass production and plant derived OM (i.e., root and harvest residues, root exudates). The output of C and N mainly derived from OM decomposition as well as leaching and solely for N from denitrification and plant uptake followed by harvest. The incorporation and distribution of plant residues and OM decomposition processes in soils are largely influenced by management options such as the tillage intensity. However, over a time span of 14 to 21 years, we found no significant differences between the C_{org} and N_t stocks (on an equivalent soil mass approach) in 0 - ~25 cm soil depth of NT, RT, and CT. In contrast, *Andruschkewitsch* et al. (2013) detected for the same study sites significantly higher C_{org} stocks for RT compared to CT and NT in 0-40 cm soil depth. In both studies, the difference between the highest (RT) and the lowest C_{org} stock (CT) is about 5 t ha^{-1} but the soil samples were taken at different times and from different points in the field. Consequently, short term and small scale heterogeneities in the distribution of soil C_{org} within the study sites might be responsible for the lacking significance in our study.

Different tillage intensities resulted in the surface soil (0 - ~5 cm) in significantly higher stocks of C_{org} and N_t under NT compared to CT (Table 4). Similar findings were described for silty loam soils by *Mikha & Rice* (2004) and *Hermle* et al. (2008). In contrast, a significant increase in C_{org} and N_t with decreasing tillage intensity was not observed for the subsurface soil (5 - ~25 cm). Main processes responsible are the different residue locations under the different tillage systems, dilution effects under CT (incorporation of the crop residues in 0 - ~25/30 cm under CT). Additionally, lower physical impact of reduced or no tillage systems compared to CT may lead to an improved physical protection of OM against microbial decomposition due to the occlusion in aggregates. Overall, our data confirm studies from *Baker* et al. (2007) and *Luo* et al. (2010) emphasizing that the C sequestration in soils under no tillage compared to conventional tillage seems to be restricted to surface soils. In greater soil depths this effect can hardly be detected.

2.4.2 Microbial biomass C and N

The stocks of C_{mic} (p ≤0.1) and N_{mic} (p ≤0.05) were higher under NT compared to CT in 0- ~5 cm soil depth (Table 4). This is in line with *Green* et al. (2007) and *Balota* et al. (2004) who also reported higher C_{mic} contents under NT compared to CT. Similar to the stocks of C_{org} and N_t no significant differences in C_{mic} and N_{mic} between the tillage systems were detected in 5 - ~25 cm soil depth. Our data indicate that the increase in C_{org} and N_t in the surface soil with decreasing tillage intensity also leads to an increase in easily decomposable OM resulting in higher stocks of C_{mic} and N_{mic}. This is supported by positive correlations between stocks of C_{org} and LF (r = 0.61, p <0.01; all soils) as well as between stocks of C_{mic} and N_{mic} and LF (r = 0.86 and r = 0.7, p <0.01; all soils).

The LF fraction as separated here represents management sensitive free and aggregate occluded organic particles (*Six* et al., 2002), which were shown to contribute to the easily decomposable OM in topsoils (*Strosser*, 2010; *Kaiser* et al., 2010). However, because of the higher level of protection against decomposition for occluded organic particles compared to free organic particles (*Six* et al., 2002), we assume the latter to be more important as an easily available energy source for microorganisms. Similar findings were reported by *Alvarez* et al. (1995), who detected a close positive relationship between the ratio of basal respiration and microbial biomass and the amount of the coarse plant debris as well as the free light fraction.

The higher macro-aggregate content detected by *Andruschkewitsch* et al. (2013) for the NT soils in 0-5 cm depth compared to CT may also contribute to the observed increase in C_{mic} and N_{mic} stocks with decreasing tillage intensity. *Balota* et al. (2004) assumed more favorable conditions for microorganisms in soils without tillage because of improved water supply due to enhanced soil aggregation compared to soils of intensive tillage. Especially macro-aggregates in topsoils act as habitats for soil microorganisms (*Bailey* et al., 2012) and can promote decomposition processes due to the close proximity of decomposers and potential energy sources (i.e., occluded OM). However, differences between size, mineral composition, and occluded OM imply that not all aggregates may host equally abundant or active microbial communities (*Bailey* et al., 2012). Consequently, macro-aggregates may act in two directions: enhanced accumulation of OM by occlusion as well as increased microbial biomass and decomposition of OM.

2.4.3 CO_2-C emission and mineralized N

The amounts of respired CO_2-C and leached N_{min} (i.e., NO_3^--N, NH_4^+-N) were significantly higher under NT compared to CT in 0- ~5 cm soil depth (Figure 1 a and b). This is confirmed for N_{min} by results from *Balesdent* et al. (2000) who compared data from eight studies and found in all cases higher mineralizable N amounts in 0-5/10 cm depth of various soil types for NT compared to CT. For the CO_2 emissions, *Oorts* et al. (2007) and *Pandey* et al. (2012) also found increased soil CO_2 losses with decreasing tillage intensity whereas *Brye* et al. (2006) and *Curtin* et al. (2000) found the opposite effect. These contradictorily results led *Soane* et al. (2012) conclude on no consistent effects of the tillage intensity on CO_2 emissions mainly because of interdependently influences of, among others, climate, soil moisture as well as amount, type, and stratification of soil OM.

The higher N_{min} and CO_2 amounts for NT compared to CT in 0- ~5 cm soil depth found in our study are in line with the above discussed increase in bio-available substrates (C_{org}, N_t, LF) and microbial biomass (C_{mic} and N_{mic}) for surface soils of NT compared to CT leading to an enhanced OM decomposition and increased CO_2 and N_{min} production. This is supported by significant correlations between stocks of C_{mic} and N_{mic} and amounts of i) CO_2-C (r = 0.80, p <0.01 and r = 0.57, p <0.01; all soils) and ii) N_{min} (r = 0.79, p <0.01 and r = 0.6, p <0.01; all soils).

2.4.4 Labile, intermediate, and passive C and N pools

We aimed to separate the bulk soil OM into pools of C and N with different turnover dynamics to analyse the effect of the tillage intensity on OM pools of different ecological functions. For example, the active or labile OM compartment (discussed below) should facilitate benefits for microbial OM turnover processes and soil productivity while the passive OM compartment (discussed below) should provide benefits to mitigate climate change.

The stocks of the modeled labile C and N pools were significantly larger under NT compared to CT (Figure 2 a and b) in 0 - ~5 cm soil depth. This is in line with the higher C_{mic}, N_{mic}, and N_{min} stocks and CO_2-C emissions for the surface NT soils compared to CT. Furthermore, correlation analyses for all soils revealed close positive relationships between the stocks of the labile C and N pools and the stocks of the LF, N_{min}, C_{mic}, CO_2-C and N_{mic} (Figure 3 a-d, Figure 4 a-d). The similar coefficients of determination for the correlations found for the labile C and the labile N pool and the close positive relationship between labile C and N_{min} as well as labile N and CO_2-C emphasize the close interaction between the fast cycling soil C and N and indicate the same source of easily bio-available OM. Our results generally confirm findings from *Hermle* et al. (2008) who investigated the effect of tillage intensity on carbon pools in a sandy loam soil under a wheat-maize-wheat-canola rotation. They found significantly higher stocks of the labile C pool under no and reduced tillage compared to conventional tillage in 0-10 cm soil depth.

Only for the intermediate N pool of the surface NT soils we found a trend of larger stocks compared to CT (Figure 2 b). With respect to the postulated MRT between 10 and 100 years for the intermediate OM pool, it takes more than 14 to 21 years of continuous management to reach steady state conditions regarding this pool. Consequently, the impact of the tillage intensity on the intermediate C and N pools cannot be seen in its full extent for the analyzed study sites. Furthermore, the intermediate OM pool seems to be stronger affected by the site characteristics than the labile pool. For example, the differences between the labile C pool of the NT and CT soils considered for each study site individually ranges from 1.1 to 1.7 t ha^{-1} (Table 2), whereas the respective differences for the intermediate C pools showed a much wider range from 0.5 to 4.1 t ha^{-1} (Table 2) emphasizing the more pronounced site specific response to differences in the tillage intensity. Similar to *Hermle* et al. (2008), the intermediate C and N pools were found to be several times larger than the labile and stable pools. In our study, the intermediate C and N pool comprises between 73% and 84% of the total C$_{org}$ and between 70% and 88% of the total N. Changes in this quantitatively most meaningful OM pool due to management might be highly important for soil ecosystem services. Based on the trend of a higher intermediate N pool in surface soils for NT compared to CT and a similar distribution pattern for the intermediate C pool (Figure 2 a and b) after 14 to 21 years we assume this differences to be more pronounced after more than 30 years. For surface soils, this would mean an increasing reservoir of OM by decreasing management intensity that is in part degradable supplying energy and nutrients. Another part of this reservoir would be stronger protected against decomposition contributing to maintain more sustainable OM levels in surface soils. The potential benefits in both directions are corroborated by positive correlations between the stocks of the intermediate C and N pools and the stocks of LF, N_{min} and C_{mic} (Figure 3 e-g, Figure 4 e-g) showing an effect of fractions mainly indicative for labile OM on the size of the intermediate pool on the hand. On the other hand, the low coefficients of determinations detected for these correlations indicate these fractions to be less important for the intermediate pool compared to the labile pool. Furthermore, the LF as separated here comprises free (i.e. easily decomposable) and aggregate occluded organic particles whereby the latter ones are stronger protected against microbial decomposition (*Six* et al., 2002) contributing to the C sequestration.

For the subsurface soil samples of 5 - ~25 cm depth, we found slightly but significantly higher labile C pools under CT compared to RT and NT, which results most likely from the deeper distribution of fresh plant litter within the CT soils (Figure 2 c). The labile N pool showed a similar trend (Figure 2 d). This is in line with the significantly larger CO_2-C and N_{min} amounts detected for the subsurface CT soils compared to NT. The intermediate C and N pools were larger under RT compared to CT and NT but the differences were not significant. The lacking significance can be partially explained by the strong influence of site specific characteristics on the intermediate pools in subsurface soils, since the differences between the intermediate C pool of the RT and CT soils considered for each study site individually ranges from 1.2 to 9.4 t ha^{-1} (Table 2). Similarly, the respective differences for the intermediate N pools also showed a wide range from -140 to 790 kg ha^{-1} (Table 2) underlining the strong site specific response to differences in the tillage intensity in subsurface soils. Overall, the data suggest a different response of labile and intermediate C and N pools to the tillage intensity in dependence of soil depth.

The experimentally determined passive C and N pools were not affected by the tillage intensity neither for the surface soils nor for the subsurface soils, which is in line with results from *Hermle* et al. (2008). The mineral characteristics of soils from the four study sites that are governed by longlasting pedogenetic processes and known to exert the main control over OM stabilization in aerated topsoils are probably important determinants for the passive C and N pools. Accordingly, the stocks of the passive C and N pools are positively correlated with the stocks of Al_α (Figure 3 h and 4 h) and clay (Figure 3 i and 4 i). The effect of clay size minerals on OM stabilization against microbial decay is well known. But in contrast to a large body of literature that showed Fe-oxides to be one of the main drivers (e.g., *Kaiser* et al., 2007; *Kleber* et al., 2005; *Mikutta* et al., 2006), we found here indications that amorphous Al-oxides (represented by Al_α) may be of particularly large importance for the passive C and N pools in the studied soils. Amorphous Al-oxides can be involved in the formation of small (<53 μm) and highly stable aggregates and/or chemically adsorb organic molecules on their surfaces (*Mikutta* et al., 2011; *Schneider* et al., 2010). Both processes are known to be effective in the stabilization of OM against microbial decomposition.

2.5 Conclusion

In this study, we aimed to analyze the impact of no-tillage, reduced tillage, and conventional tillage on OM pools of different turnover dynamics in surface soils compared to subsurface soils.

In surface soils, only no tillage resulted in larger stocks of the labile C and N pool compared to conventional tillage in surface soils after 14 to 22 years. The increase of labile pools in surface soils due to decreasing tillage intensity should have positive effects for the nutrient cycling and productivity, which is especially important for the development of more sustainable managed agro-ecosystems. However, these effects were not detected in subsurface soils indicating different effects of the tillage intensity on the OM dynamics with increasing soil depth. Because of the deeper distribution of plant residues, the stocks of the labile C and N pool in subsurface soils were larger under conventional tillage compared to reduced or no tillage.

The intermediate C and N pools represented between 70 and 88% of the total C_{org} and N and are quantitatively much more important for the OM dynamics than the passive or labile pools. Only the intermediate N pool in the surface soils was increased by decreasing tillage intensity with potential positive effects for the soil productivity but the magnitude of such increase seems to be highly site specific.

The passive C and N pools were not affected by the tillage intensity neither in surface soils nor in subsurface soils. Here, mineral characteristics such as the amount and the composition of the clay size fraction seem to be main driving factors.

2.6 References

Alvarez, R., Santanatoglia, O. J., Daniel, P. E., Garcia, R. (1995): Respiration and specific activity of soil microbial biomass under conventional and reduced tillage. *Pesquisa Agropecuaria Brasileira* 30, 701–709.

Andruschkewitsch, R., Geisseler, D., Koch, H.-J., Ludwig, B. (2013): Effects of tillage on contents of organic carbon, nitrogen, water-stable aggregates and light fraction for four different long-term trials. *Geoderma* 192, 368–377.

Bailey, V. L., Smith, J. L., Bolton, H. (2002): Fungal-to-bacterial ratios in soils investigated for enhanced C sequestration. *Soil Biology & Biochemistry* 34, 997–1007.

Baker, J. M., Ochsner, T. E., Venterea, R. T., Griffis, T. J. (2007): Tillage and soil carbon sequestration - What do we really know? *Agriculture Ecosystems & Environment* 118, 1–5.

Balesdent, J., Chenu, C., Balabane, M. (2000): Relationship of soil organic matter dynamics to physical protection and tillage. *Soil & Tillage Research* 53, 215–230.

Balesdent, J., Pétraud J. P., Feller C. (1991): Effets des ultrasons sur la distribution granulométrique des matières organiques des sols. *Science du Sol* 29, 95–106.

Balota, E. L., Colozzi, A., Andrade, D. S., Dick, R. P. (2004): Long-term tillage and crop rotation effects on microbial biomass and C and N mineralization in a Brazilian Oxisol. *Soil & Tillage Research* 77, 137–145.

Brookes, P. C., Landman, A., Pruden, G., Jenkinson, D. S. (1985): Chloroform fumigation and the release of soil-nitrogen - a rapid direct extraction method to measure microbial biomass nitrogen in soil. *Soil Biology & Biochemistry* 17, 837–842.

2 Impact of tillage intensity on carbon and nitrogen pools in surface and subsurface soils of four long-term field experiments

Brye, K. R., Longer, D. E., Gbur, E. E. (2006): Impact of tillage and residue burning on carbon dioxide flux in a wheat-soybean production system. *Soil Science Society of America Journal* 70, 1145–1154.

Cambardella, A. C. A., Elliott, E. T. (1993): Carbon and nitrogen distribution in aggregates from cultivated and native grassland soils. *Soil Science Society of America Journal* 57, 1071–1076.

Coppens, F., Merckx, R., Recous, S. (2006): Impact of crop residue location on carbon and nitrogen distribution in soil and in water-stable aggregates. *European Journal of Soil Science* 57, 570–582.

Curtin, D., Wang, H., Selles, F., McConkey, B. G., Campbell, C. A. (2000): Tillage effects on carbon fluxes in continuous wheat and fallow-wheat rotations. *Soil Science Society of America Journal* 64, 2080–2086.

DIN Deutsches Institut für Normung e.V. (1997): Bodenuntersuchungsverfahren im Landwirtschaftlichen Wasserbau - Chemische Laboruntersuchungen - Teil 6: Bestimmung des Gehaltes an oxalatlöslichem Eisen: DIN 19684-6:1997-12. Beuth Verlag, Berlin.

DIN Deutsches Institut für Normung e.V. (2001): Bodenbeschaffenheit - Bestimmung der Trockenrohdichte: DIN ISO 11272:2001-01. Beuth Verlag, Berlin.

DIN Deutsches Institut für Normung e.V. (2002): Bodenbeschaffenheit - Bestimmung der Partikelgrößenverteilung in Mineralböden - Verfahren mittels Siebung und Sedimentation: DIN ISO 11277:2002-08. Beuth Verlag, Berlin.

DIN Deutsches Institut für Normung e.V. (2009): Bodenbeschaffenheit - Felduntersuchungen - Teil 13: Bestimmung der Carbonate, der Sulfide, des pH-Wertes und der Eisen(II)-Ionen: DIN 19682-13:2009-01. Beuth Verlag, Berlin.

Ellert, B. H., Bettany, JR (1995): Calculation of organic matter and nutrients stored in soils under contrasting management regimes. *Canadian Journal of Soil Science* 75, 529–538.

Fierer, N., Schimel, J. P. (2002): Effects of drying-rewetting frequency on soil carbon and nitrogen transformations. *Soil Biology & Biochemistry* 34, 777–787.

Green, V. S., Stott, D. E., Cruz, J. C., Curi, N. (2007): Tillage impacts on soil biological activity and aggregation in a Brazilian Cerrado Oxisol. *Soil & Tillage Research* 92, 114–121.

Heitkamp, F., Raupp, J., Ludwig, B. (2009): Impact of fertilizer type and rate on carbon and nitrogen pools in a sandy Cambisol. *Plant and Soil* 319, 259–275.

Helfrich, M., Flessa, H., Mikutta, R., Dreves, A., Ludwig, B. (2007): Comparison of chemical fractionation methods for isolating stable soil organic carbon pools. *European Journal of Soil Science* 58, 1316–1329.

Hermle, S., Anken, T., Leifeld, J., Weisskopf, P. (2008): The effect of the tillage system on soil organic carbon content under moist, cold-temperate conditions. *Soil & Tillage Research* 98, 94–105.

Jacobs, A., Helfrich, M., Hanisch, S., Quendt, U., Rauber, R., Ludwig, B. (2010): Effect of conventional and minimum tillage on physical and biochemical stabilization of soil organic matter. *Biology and Fertility of Soils* 46, 671–680.

Jagadamma, S., Lal, R. (2010): Integrating physical and chemical methods for isolating stable soil organic carbon. *Geoderma* 158, 322–330.

Janzen, H. H. (2004): Carbon cycling in earth systems - a soil science perspective. *Agriculture Ecosystems & Environment* 104, 399–417.

Joergensen, R. G. (1996): The fumigation-extraction method to estimate soil microbial biomass: Calibration of the k(EC) value. *Soil Biology & Biochemistry* 28, 25–31.

Kader, M. A., Sleutel, S., Begum, S. A., D'Haene, K., Jegajeevagan, K., Neve, S. de (2010): Soil organic matter fractionation as a tool for predicting nitrogen mineralization in silty arable soils. *Soil Use and Management* 26, 494–507.

Kaiser, K., Mikutta, R., Guggenberger, G. (2007): Increased stability of organic matter sorbed to ferrihydrite and goethite on aging. *Soil Science Society of America Journal* 71, 711–719.

Kaiser, M., Ellerbrock, R. H., Wulf, M., Dultz, S., Hierath, C., Sommer, M. (2012): The influence of mineral characteristics on organic matter content, composition, and stability of topsoils under long-term arable and forest land use. *Journal of Geophysical Research-Biogeosciences* 117, 16 pages.

Kaiser, M., Wirth, S., Ellerbrock, R. H., Sommer, M. (2010): Microbial respiration activities related to sequentially separated, particulate and water-soluble organic matter fractions from arable and forest topsoils. *Soil Biology & Biochemistry* 42, 418–428.

Kladivko, E. J. (2001): Tillage systems and soil ecology. *Soil & Tillage Research* 61, 61–76.

Kleber, M., Mikutta, R., Torn, M. S., Jahn, R. (2005): Poorly crystalline mineral phases protect organic matter in acid subsoil horizons. *European Journal of Soil Science* 56, 717–725.

Kögel-Knabner, I., Guggenberger, G., Kleber, M., Kandeler, E., Kalbitz, K., Scheu, S., Eusterhues, K., Leinweber, P. (2008): Organo-mineral associations in temperate soils: Integrating biology, mineralogy, and organic matter chemistry. *Journal of Plant Nutrition and Soil Science* 171, 61–82.

Luo, Z., Wang, E., Sun, O. J. (2010): Can no-tillage stimulate carbon sequestration in agricultural soils? A meta-analysis of paired experiments. *Agriculture Ecosystems & Environment* 139, 224–231.

von Lützow, M., Kögel-Knabner, I., Ekschmitt, K., Flessa, H., Guggenberger, G., Matzner, E., Marschner, B. (2007): SOM fractionation methods: Relevance to functional pools and to stabilization mechanisms. *Soil Biology & Biochemistry* 39, 2183–2207.

von Lützow, M., Kögel-Knabner, I., Ekschmitt, K., Matzner, E., Guggenberger, G., Marschner, B., Flessa, H. (2006): Stabilization of organic matter in temperate soils: mechanisms and their relevance under different soil conditions - a review. *European Journal of Soil Science* 57, 426–445.

Mikha, M. M., Rice, C. W. (2004): Tillage and manure effects on soil and aggregate-associated carbon and nitrogen. *Soil Science Society of America Journal* 68, 809–816.

Mikutta, R., Kleber, M., Torn, M. S., Jahn, R. (2006): Stabilization of soil organic matter: Association with minerals or chemical recalcitrance? *Biogeochemistry* 77, 25–56.

Mikutta, R., Zang, U., Chorover, J., Haumaier, L., Kalbitz, K. (2011): Stabilization of extracellular polymeric substances (Bacillus subtilis) by adsorption to and coprecipitation with Al forms. *Geochimica et Cosmochimica Acta* 75, 3135–3154.

Oorts, K., Garnier, P., Findeling, A., Mary, B., Richard, G., Nicolardot, B. (2007): Modeling soil carbon and nitrogen dynamics in no-till and conventional tillage using PASTIS model. *Soil Science Society of America Journal* 71, 336–346.

Pandey, D., Agrawal, M., Bohra, J. S. (2012): Greenhouse gas emissions from rice crop with different tillage permutations in rice-wheat system. *Agriculture Ecosystems & Environment* 159, 133–144.

R Development Core Team (2010): R: A language and environment for statistical computing.

Salinas-Garcia, J. R., Hons, F. M., Matocha, J. E., Zuberer, D. A. (1997): Soil carbon and nitrogen dynamics as affected by long-term tillage and nitrogen fertilization. *Biology and Fertility of Soils* 25, 182–188.

Schmidt, M. W. I., Torn, M. S., Abiven, S., Dittmar, T., Guggenberger, G., Janssens, I. A., Kleber, M., Kögel-Knabner, I., Lehmann, J., Manning, D. A. C., Nannipieri, P., Rasse, D. P., Weiner, S., Trumbore, S. E. (2011): Persistence of soil organic matter as an ecosystem property. *Nature* 478, 49–56.

Schneider, M. P. W., Scheel, T., Mikutta, R., van Hees, P., Kaiser, K., Kalbitz, K. (2010): Sorptive stabilization of organic matter by amorphous Al hydroxide. *Geochimica et Cosmochimica Acta* 74, 1606–1619.

Six, J., Conant, R. T., Paul, E. A., Paustian, K. (2002): Stabilization mechanisms of soil organic matter: Implications for C-saturation of soils. *Plant and Soil* 241, 155–176.

Six, J., Elliott, E. T., Paustian, K. (2000): Soil macroaggregate turnover and microaggregate formation: a mechanism for C sequestration under no-tillage agriculture. *Soil Biology & Biochemistry* 32, 2099–2103.

Soane, B. D., Ball, B. C., Arvidsson, J., Basch, G., Moreno, F., Roger-Estrade, J. (2012): No-till in northern, western and south-western Europe: A review of problems and opportunities for crop production and the environment. *Soil & Tillage Research* 118, 66–87.

Stanford, G., Smith, S. J. (1972): Nitrogen Mineralization Potentials of Soils. *Soil Science Society of America Journal* 36, 465–472.

Strosser, E. (2010): Methods for determination of labile soil organic matter: An overview. *Journal of Agrobiology* 27, 49–60.

Tisdall, J. M., Oades, J. M. (1982): Organic-Matter and Water-Stable Aggregates in Soils. *Journal of Soil Science* 33, 141–163.

Vance, E. D., Brookes, P. C., Jenkinson, D. S. (1987): An Extraction Method for Measuring Soil Microbial Biomass-C. *Soil Biology & Biochemistry* 19, 703–707.

Wang, W. J., Smith, C. J., Chen, D. (2003): Towards a standardised procedure for determining the potentially mineralisable nitrogen of soil. *Biology and Fertility of Soils* 37, 362–374.

2   Impact of tillage intensity on carbon and nitrogen pools in surface and subsurface soils of four long-term field experiments

Watts, D. B., Torbert, H. A., Prior, S. A., Huluka, G. (2010): Long-Term Tillage and Poultry Litter Impacts Soil Carbon and Nitrogen Mineralization and Fertility. *Soil Science Society of America Journal* 74, 1239–1247.

Zibilske, L. M., Bradford, J. M., Smart, JR (2002): Conservation tillage induced changes in organic carbon, total nitrogen and available phosphorus in a semi-arid alkaline subtropical soil. *Soil & Tillage Research* 66, 153–163.

Zotarelli, L., Alves, B. J. R., Urquiaga, S., Boddey, R. M., Six, J. (2007): Impact of tillage and crop rotation on light fraction and intra-aggregate soil organic matter in two Oxisols. *Soil & Tillage Research* 95, 196–206.

3 Long-term tillage effects on the distribution of P fractions of German loess soils

Christiane Piegholdt[1], Daniel Geisseler[1], Heinz-Josef Koch[2], Bernard Ludwig[1*]

[1]*Department of Environmental Chemistry, University of Kassel, Nordbahnhofstr. 1a, 37213 Witzenhausen, Germany*

[2]*Institute of Sugar Beet Research, Holtenser Landstr. 77, 37079 Göttingen, Germany*

corresponding author: bludwig@uni-kassel.de

Key words: P fractionation, conventional tillage, no-till, inorganic and organic P pools, P stocks

Abstract

Different tillage systems may affect phosphorus (P) dynamics in soils due to differently distributed plant residues, different aggregate dynamics and erosion losses, but quantitative data are scarce. Objectives were to investigate the effect of tillage on the availability of P in a long-term field trial on loess soils (Phaeozems and Luvisols) initiated from 1990 to 1997. Four research sites in east and south Germany were established with a crop rotation consisting of two times winter wheat followed by sugar beet. The treatments were no-till (NT) without cultivation, except for seedbed preparation to a depth of 5 cm before sugar beet was sown and conventional tillage (CT) with mouldboard ploughing down to 25-30 cm. Soil P was divided into different pools by a sequential extraction method and total P (P_t) in the single P fractions was extracted by digesting the extracts of the fractionation to calculate the contents of organic P. The P_t content (792 mg kg^{-1} soil) in the topsoil (0-5 cm) of NT was 15% higher compared to CT, while with increasing depth the P_t content decreased more under NT than under CT. This

3 Long-term tillage effects on the distribution of P fractions of German loess soils

was also true for the other P fractions except for residual P. The higher P contents in the topsoil of NT presumably resulted from the shallower incorporation of harvest residues and fertilizer P compared to CT, whereas estimated soil losses and thus also P losses due to water erosion were only small for both treatments. Contents of oxalate extractable iron and organic carbon were positively related to the labile fractions of inorganic P, while there was a high correlation of the stable fractions with the clay contents and pH. Multiple regression analyses explained 50% of the variability of these P fractions. Overall, only small differences in the P fractions and availability were observed between the long-term tillage treatments.

3.1 Introduction

Phosphorus (P) is an essential nutrient for all organisms. In soil, P exists in many different organic and inorganic forms. Most of the inorganic P (P_i) is adsorbed to mineral surfaces or precipitated as secondary P compounds so that only a small part of P_i is plant available (*Schmidt* et al., 1996; *Redel* et al., 2007). The pH is an important factor for the availability of P and inorganic forms of P in soils. For instance, *Köster & Nieder* (2007) reported that in sandy soils with a pH around 5.5, P is mostly bound at Fe and Al oxides, whereas in loess soils with a pH around 6.5, P is primary bound as Ca phosphates.

In order to adjust P fertilization to plant demand, information about the availability of P is required. In agricultural systems, crop management has a strong effect on the soil P content and P availability. Tillage influences physical and chemical processes in soil (*Hedley* et al., 1982; *Vu* et al., 2009) and may therefore also affect the retention and stabilization of P (*Wright*, 2009; *Messiga* et al., 2011). Intensive tillage may decrease soil organic matter content and soil structure compared to reduced tillage (*D'Haene* et al., 2008), potentially reducing organically bound P (P_O) and resulting in increased P losses through different aggregate dynamics, erosion and runoff (*Scholz* et al.,

2008). Tillage also mixes the soil to a certain depth, preventing a stratification within the tilled soil layer (*Vu* et al., 2009). The effect of tillage on the distribution of P in the soil profile has been studied in detail, whereas its effect on the different P pools has received less attention (*O'Halloran*, 1993; *Redel* et al., 2007; *Vu* et al., 2009; *Deubel* et al., 2011; *Messiga* et al., 2011).

Tillage induced effects on P pools may be studied using a sequential P fractionation procedure, which differentiates between P pools according to their solubility in different extractants (*Srivastava & Pathak*, 1971; *Hedley* et al., 1982; *Guo* et al., 2000). *O'Halloran* (1993) applied a sequential extraction procedure to Canadian soils, a clayey Gleysol and a sandy loamy Brunisol, under different soil management to differentiate between P pools according to their availability for plants. He found that the tilling intensity and depth affected labile and moderately labile P pools with higher contents under no-till (NT) compared to conventional tillage (CT). *Redel* et al. (2007) studied the effects of tillage and crop rotation on P pools in a Chilean Ultisol and reported that NT led to a nutrient stratification that benefitted shallower rooted crops. The combined effect of an adoption of NT and low-chemical input systems was compared with a conventional (CT and mineral fertilization) system by *Daroub* et al. (2001). They reported that adoption of NT and low-chemical input systems with a winter leguminous cover crop in the rotation for seven years did not increase organic P significantly in any of the fractions extracted from the annual cropping systems studied. In contrast, continuous alfalfa resulted in an increase of organic P extracted by NaOH.

Overall, the studies above indicate that tillage systems and crops may affect the distribution of P in different pools. Moreover, climate and relief (P losses through erosion), soil texture (affecting erosion and also the extent P adsorption) (*Sathya* et al., 2009) and the amount and type of P fertilizer used (*Shafquat & Pierzynski*, 2010) may affect the P dynamics.

3 Long-term tillage effects on the distribution of P fractions of German loess soils

In order to get an improved understanding on the P dynamics in loess soils affected by CT and NT, we used the information of four long-term trials and set ourselves the following objectives: (i) to determine the effects of long-term tillage practices on P availability and the distribution of organic and inorganic P fractions in the soil profiles; and (ii) to study the relationship between P fractions and soil texture and selected chemical soil properties.

3.2 Material and methods

3.2.1 Experimental sites and treatments

A long-term field trial was initiated at four sites (Friemar, Grombach, Lüttewitz and Zschortau) in east and south Germany between 1990 and 1997 by the Institute of Sugar Beet Research in cooperation with the agricultural division of Südzucker AG (Table 4). The annual mean air temperature at the four sites ranges from 8.0 to 9.3 °C and the annual precipitation from 512 to 776 mm (data were provided by Deutscher Wetterdienst). Soil texture varies considerably between the four loess sites with clay contents ranging from 15% (Zschortau) to 30% (Friemar). The sand content is below 6%, except for Zschortau, where it reaches 26% (Table 4). At all sites, the crop rotation includes two growing seasons of winter wheat (*Triticum aestivum* L.), followed by sugar beet (*Beta vulgaris* ssp. *vulgaris* var. *altissima* DÖLL). Crop residues are left in the field. At Friemar and Grombach, winter wheat was sown in fall 2009 after the sugar beet harvest, while at Lüttewitz und Zschortau, winter wheat was sown after previous wheat.

The trial consists of the treatments CT and NT. At each location a field has been divided into strips, each ranging from 2.5 to 8 ha. The strip under CT includes mouldboard ploughing to a depth of 25-30 cm, the NT strip involves no cultivation, except for seedbed preparation with a rigid-tine cultivator or disc harrow to a depth of

3 Long-term tillage effects on the distribution of P fractions of German loess soils

5 cm before the sugar beets are sown. More detailed information on the sites is given by *Dieckmann* et al. (2006) and *Koch* et al. (2009).

Mineral fertilizers (N, P, K and Mg) as well as calcium carbonate (lime) were applied according to site-specific recommendations based on soil analyses. Fertilization was identical for the two tillage treatments at each site and year. Nitrogen fertilizers were applied annually according to the results of the EUF soil analyses (*Mengel & Kirkby*, 2001). Phosphorus, Mg, S and K fertilizers were also given according to EUF recommendation and differed between sites and years, but were uniform for the tillage treatments at one site in one year (*Koch* et al., 2009).

Phosphorus fertilizers (triple superphosphate ($Ca(H_2PO_4)_2 \cdot H_2O$) and diammoniumhydrogenphosphate ($(NH_4)_2HPO_4$) were applied from 2003 to 2008 and rates ranged from 5 to 23 (Friemar), 11 to 32 (Grombach), 5 to 15 (Lüttewitz), 4 to 7 kg P (ha year)$^{-1}$ (Zschortau). The last P fertilization was at Friemar in 2008 (5 kg P ha^{-1}), at Grombach in 2008 (18 kg P ha^{-1}), at Lüttewitz in 2006 (5 kg P ha^{-1}) and at Zschortau in 2007 (4 kg P ha^{-1}).

Table 4: Climatic conditions (long-term annual means) and soil properties at the four research sites from the four long-term field experiments. Values shown for texture are means of the three pseudoreplicates per tillage system and study site.

Site	Site characteristics				Soil properties			Soil type (WRB, 2006)
	Year established	Altitude (m)	Temperature (°C)	Precipitation (mm)	Sand	Silt	Clay	
					(g kg^{-1})			
Friemar (Thuringia)	1992/1993	310	8.0	554	54	645	301	Haplic Phaeozem
Grombach (Baden-Wuerttemberg)	1990/1991	270	9.3	776	18	727	255	Luvisol/ Phaeozem
Lüttewitz (Saxony)	1992/1993	290	8.6	572	25	799	176	Luvisol
Zschortau (Saxony)	1997/1998	110	8.8	512	264	583	153	Haplic Luvisol /Haplic Planosol

3.2.2 Soil sampling and analyses

Soil samples were taken in April 2010 at all four locations from the long-term field trial. For soil sampling, each tillage strip was divided into three plots along the length of the field and a composite sample consisting of five cores (4 cm diameter) was taken from each plot at three depths (0-5, 5-25 and 25-40 cm). All soil samples were sieved (≤2 mm) and stored field moist at 4 °C.

Field moist soil samples were analyzed for pH (2.5 ml (g soil)$^{-1}$, 0.01 M $CaCl_2$), dry samples were used to determine texture using the pipet method (*Köhn*, 1928) and total carbon (C) and nitrogen (N) content by dry combustion (Elementar Vario El, Heraeus, Hanau, Germany). Carbonate-C (CO_3-C) was determined by measuring the increase in pressure after the addition of 20 ml of 10% HCl to a soil sample (5-10 g). Gravimetric soil moisture content was determined by drying samples at 105 °C for 24 h. Bulk density was determined according to *DIN ISO 11272* (1998).

For the determination of oxalate extractable iron (Fe_{ox}) and aluminium (Al_{ox}), following *DIN 19684-6* (1997), a 5 g sample was shaken for 2 hours with 50 ml extraction solution (0.1 M ammoniumoxalate and 0.1 M oxalic acid). After filtration through a fibreglass filter, the Al_{ox}- and Fe_{ox}-concentrations in the filtrates were measured by using an atomic absorption spectrometer (Al_{ox}: Model GBC 906 AAS, GBC Scientific Equipment, Braeside, Australia; Fe_{ox}: Unicam 939 AAS with a Gilson 222 Rack 22 autosampler, Villiers, France).

3.2.3 Phosphorus fractionation

Field-moist soil samples were used to determine the different P fractions following the sequential extraction method by *Hedley* et al. (1982) with modifications described by *Tiessen & Moir* (1993). Briefly, to 0.65 g of field moist soil, 40 ml of the respective extractant were added and the suspensions were shaken for 16 h. The suspensions were

3   Long-term tillage effects on the distribution of P fractions of German loess soils

then centrifuged at 4000 xg for 25 min, filtered through folded filter paper (Whatman no. 595 ½) and stored in a freezer until analyzed for P.

In a first step, labile P was extracted with DI water (DI-P), followed by extraction with 0.5 M NaHCO$_3$ (NaHCO$_3$-P$_i$). These extractants extract the labile, plant available P present in the soil solution or adsorbed on surfaces of more crystalline P compounds, sesquioxides or carbonates (*Mattingly*, 1975; *Bowman & Cole*, 1978; *Tiessen & Moir*, 1993; *Guo* et al., 2000; *Frizano* et al., 2002; *Redel* et al., 2007).

Phosphorus adsorbed to clay minerals and associated with amorphous and some crystalline Fe and Al phosphates (*Tiessen* et al., 1984) were extracted with 0.1 M NaOH (NaOH-P$_i$) followed by 1 M HCl (HCl-P$_i$), which extracted P largely bound to Ca (*Hedley* et al., 1982; *Tiessen & Moir*, 1993; *Guo* et al., 2000; *Frizano* et al., 2002; *Redel* et al., 2007). These two fractions are moderately labile and are considered to be long-term plant available through soil P transformations (*Guo & Yost*, 1998; *Guo* et al., 2000; *Frizano* et al., 2002).

In a last step, residual-P, present in the form of stable Al- and Fe-phosphates was extracted by digesting the residual soil sample with H$_2$SO$_4$ and H$_2$O$_2$ (*Hedley* et al., 1982; *Tiessen & Moir*, 1993; *Cross & Schlesinger*, 1994). For the digestion, the soil sample was transferred into Teflon vials with 5 ml of concentrated H$_2$SO$_4$ and 2 ml of 30% H$_2$O$_2$ were added. The suspension was then digested in the microwave for 20 min at 210 °C. After cooling to 50 °C, the suspension was filtered through folded filterpaper (Whatman no. 595 ½). The solution was adjusted to pH 10 with 10 M NaOH using p-nitrophenol as pH-indicator. The volume was then adjusted to 70 ml with DI water.

Soil P$_O$ was determined indirectly in the NaHCO3 (NaHCO3-Pt) and NaOH (NaOH-Pt) extracts according to the method developed by *Bowman & Moir* (1993) with slight modifications. Briefly, an aliquot of 4 ml of the extract was mixed with 0.5 g of potassium peroxide disulfate (K$_2$S$_2$O$_8$) and 2 ml of 5.5 M H$_2$SO4 and digested in the microwave for 15 min at 180 °C. The solution was then adjusted to pH 10 with 5 M NaOH as described above and the volume adjusted to 25 ml with DI water. Soil P_o was calculated as the difference between the digested and the undigested sample of the same fraction (NaHCO$_3$-P$_o$ and NaOH-P$_o$, respectively). The P concentration in the different extracts was determined by the ascorbic acid method (*Murphy & Riley*, 1962) and absorbance was measured colorimetrically with a photospectrometer (FLUOstar Omega Microplate Reader, BMG LABTECH, Ortenberg, Germany). Total P was calculated as the sum of all fractions. P stocks were calculated for the depth ranges 0-5 cm, 5-25 cm and 25-40 cm using the measured P contents, bulk densities and considered depth ranges. P stocks were not calculated on an equivalent soil mass, since bulk densities were not significantly different between tillage treatments.

3.2.4 CAL extractable Phosphorus

CAL-extractable P (CAL-P) was determined as described in *Schüller* (1969). Briefly, 2 g field moist soil was shaken for 90 min with a calcium-acetate-lactate-solution (0.1 M Ca lactate, 0.1 M Ca acetate, 0.3 M acetic acid, pH 4.1) and filtered through folded filters (Whatman No. 595 ½). The P contents in the filtrates were determined by the ascorbic acid method (*Murphy & Riley*, 1962). The CAL method extracts mainly Ca bound phosphates (*Köster & Nieder*, 2007).

3.2.5 Statistical analysis and potential erosion losses

The data was analyzed as a split-plot design with tillage treatment as the main factor and soil depth as sub-factor. Analysis of variance was performed on the average values of the three plots. The four sites served as field replicates. For all properties, the site * treatment and site * depth interactions were not significant ($p > 0.05$).

For correlation analyses, the three plots per tillage strip were used separately. Data were statistically analyzed with R 2.11.1 (*R Development Core Team*, 2010). Effects were considered to be significant at p ≤0.05.

Multiple regression analyses were conducted with SAS (*SAS Institute*, 2010). The influence of selected soil properties (clay, Fe_{ox}, Al_{ox}, C_{org}, CO_3-C and pH) on the different P fractions were analyzed by multiple regression analysis with stepwise variable selection. The model with the lowest Mallow's Cp was selected.

For the estimation of potential soil losses by water erosion, the parameters required were precipitation erodibility (estimated using the texture classes listed in KA 5, *AG Boden*, 2005), measure of the productivity of agricultural land, hillside length, slope and tillage and cropping system (*Schwertmann* et al., 1987). Annual precipitation ranged from 512 to 776 mm (Table 4), the measure of the productivity of agricultural land ranged from 66 to 80. The hillside length was in the range of 590 to 660 m at Friemar, Lüttewitz and Zschortau, whereas it was only 350 at Grombach. In contrast, the slope was markedly higher at Grombach (5.5%) and Lüttewitz (5%) compared to Friemar (1.5%) and Zschortau (0.25%) (*Koch, H.-J.*; unpublished data).

3.3 Results and discussion

3.3.1 Soil characteristics and control factors for P availability

Some soil characteristics (pH, contents of clay, C_{org}, carbonate, Al_{ox} and Fe_{ox}) are known as control factors on P contents and availability. The pH averaged 6.9 and was not significantly affected by sampling depth and tillage treatment (Table 6). Clay contents increased in the order Zschortau (15%) < Lüttewitz (18%) < Grombach (26%) < Friemar (30%, Table 4). Contents of C_{org} decreased significantly with increasing soil depth under NT (17.8 g kg^{-1} in the top 5 cm, 6.2 g kg^{-1} between 25 and 40 cm), whereas under CT the decrease with depth (from 11.1 to 8.1 g kg^{-1}) was much less pronounced (Table 6). Less than 10% of the total C was in the form of carbonates. Contents of Fe_{ox} and Al_{ox} were not significantly affected by the tillage treatment (Table 6). Overall, the four sites comprised a wide range of C contents and different texture composition which are expected to affect P dynamics in soils and were studied in the correlation analyses described below.

Table 5: Crop yields and Calcium-Acetate-Lactate extractable P (CAL-P) status at the experimental sites under conventional tillage (CT) and no-till (NT). Values shown are means and standard errors of the mean (winter wheat: n = 5, sugar beet: n = 2 (means for 5 and 2 years, respectively), CAL-P status n = 3 (means for 3 pseudoreplicates for each field)).

Site	Treatment	Grain and taproot yields [a] (averaged 2004-2010)		CAL-P status (mg kg^{-1})			
		Winter wheat	Sugar beet	(0-25 cm)	(0-5 cm)	(5-25 cm)	(25-40 cm)
		(Mg ha^{-1})					
Friemar (Thuringia)	CT	8.2 (0.5)	74.3 (1.4)	20.9 (3.7)	22.8 (4.9)	20.4 (3.4)	15.1 (2.7)
	NT	7.6 (0.7)	69.8 (3.2)	19.6 (3.2)	36.8 (6.6)	15.3 (2.8)	3.4 (0.2)
Grombach (Baden-Wuerttemberg)	CT	7.3 (0.5)	69.9 (0.3)	14.5 (1.4)	16.2 (2.7)	14.1 (1.1)	10.1 (1.2)
	NT	7.3 (0.5)	58.5 (0.4)	17.2 (1.1)	26.4 (1.0)	14.9 (1.2)	3.9 (1.3)
Lüttewitz (Saxony)	CT	8.8 (0.7)	67.8 (2.9)	42.3 (13.2)	44.7 (12.9)	41.7 (13.3)	36.6 (11.0)
	NT	8.3 (0.6)	61.1 (3.4)	40.1 (7.5)	65.8 (4.5)	33.7 (8.8)	22.9 (3.0)
Zschortau (Saxony)	CT	8.1 (0.9)	62.3 (4.2)	45.2 (2.4)	46.4 (5.4)	44.9 (1.7)	33.3 (2.4)
	NT	7.8 (1)	56.6 (0.2)	57.9 (5.1)	57.8 (3.4)	57.9 (5.5)	34.7 (7.0)

[a] Yields are given in dry matter for winter wheat (including 14% water content) and in taproot fresh matter for sugar beet

3 Long-term tillage effects on the distribution of P fractions in German loess soils

Table 6: Contents of organic and carbonate-C (C$_{org}$, CO$_3$-C), total N (N$_t$), oxalat soluble Al and Fe (Al$_{ox}$, Fe$_{ox}$), as well as pH in the soil profile under conventional tillage (CT) and no-till (NT). Values shown are for means and standard errors for the individual sites (n = 3 pseudoreplicates) and means and standard errors of the sites (n = 4 field replicates). Values followed by the same letter are not significantly different (p ≤0.05). Capital letters refer to the comparison of tillage treatments within one depth, while lower case letters refer to sampling depth within one tillage treatment.

Soil property													
		pH		C$_{org}$		CO$_3$-C		N$_t$		Fe$_{ox}$		Al$_{ox}$	
Tillage treatment		CT	NT	CT	NT	CT	NT	CT	NT	CT	NT	CT	NT
Site	Soil depth (cm)												
Frie-mar	0-5	7.0 (0.2)	7.3 (0.0)	13.9 (0.7)	21.4 (0.9)	0.14 (0.05)	0.35 (0.07)	1.49 (0.06)	2.20 (0.05)	1.1 (0.1)	1.2 (0.1)	0.78 (0.07)	0.75 (0.08)
	5-25	7.1 (0.1)	7.1 (0.1)	14.1 (1.1)	11.9 (1.9)	0.07 (0.03)	0.17 (0.10)	1.44 (0.08)	1.23 (0.18)	1.1 (0.0)	1.3 (0.1)	0.8 (0.09)	0.87 (0.08)
	25-40	7.1 (0.2)	7.0 (0.1)	10.6 (0.8)	7.8 (0.8)	0.08 (0.02)	0.11 (0.10)	1.09 (0.07)	0.82 (0.06)	1.1 (0.1)	1.3 (0.2)	0.84 (0.08)	0.88 (0.10)
Grom-bach	0-5	6.4 (0.5)	5.8 (0.3)	8.9 (1.0)	16.2 (0.6)	2.15 (2.12)	0.02 (0.01)	0.99 (0.11)	1.67 (0.02)	1.2 (0.3)	2.0 (0.0)	0.63 (0.11)	0.71 (0.03)
	5-25	6.5 (0.4)	6.2 (0.1)	8.3 (0.1)	9.3 (0.4)	2.47 (2.46)	0.03 (0.01)	0.92 (0.06)	1.06 (0.04)	1.3 (0.4)	2.1 (0.1)	0.63 (0.11)	0.77 (0.01)
	25-40	6.8 (0.3)	6.6 (0.0)	5.3 (0.4)	4.4 (0.6)	4.91 (4.89)	0.02 (0.01)	0.59 (0.03)	0.57 (0.05)	1.1 (0.4)	1.8 (0.0)	0.65 (0.15)	0.84 (0.03)
Lütte-witz	0-5	6.9 (0.1)	6.7 (0.2)	11.0 (1.2)	17.7 (1.0)	0.07 (0.02)	0.18 (0.07)	1.26 (0.08)	1.89 (0.06)	2.1 (0.1)	1.9 (0.1)	0.57 (0.04)	0.49 (0.04)
	5-25	6.9 (0.1)	6.9 (0.2)	10.6 (1.0)	10.4 (1.1)	0.05 (0.04)	0.08 (0.04)	1.22 (0.08)	1.19 (0.06)	2.2 (0.1)	2.0 (0.1)	0.57 (0.06)	0.56 (0.04)
	25-40	7.0 (0.1)	6.9 (0.1)	8.8 (1.2)	6.5 (0.7)	0.05 (0.03)	0.04 (0.01)	1.00 (0.10)	0.76 (0.07)	2.2 (0.1)	2.2 (0.1)	0.57 (0.06)	0.6 (0.09)
Zschor-tau	0-5	7.1 (0.2)	6.9 (0.1)	10.5 (0.3)	16.0 (0.4)	0.10 (0.05)	0.24 (0.07)	1.07 (0.05)	1.55 (0.04)	1.7 (0.3)	1.6 (0.2)	0.61 (0.09)	0.53 (0.05)
	5-25	7.2 (0.1)	7.2 (0.0)	10.7 (0.4)	11.0 (0.3)	0.17 (0.07)	0.12 (0.01)	1.07 (0.04)	1.07 (0.04)	1.6 (0.3)	1.6 (0.2)	0.64 (0.09)	0.58 (0.05)
	25-40	7.3 (0.1)	7.1 (0.0)	7.7 (0.3)	6.1 (0.6)	0.26 (0.12)	0.03 (0.03)	0.78 (0.07)	0.62 (0.06)	1.9 (0.4)	1.7 (0.3)	0.65 (0.08)	0.64 (0.08)
Means of the sites	0-5	6.9 (0.2) a	6.7 (0.3)	11.1 (1.1) Aa	17.8 (1.3) Ba	0.61 (0.50)	0.20 (0.07) a	1.20 (0.11) Aa	1.83 (0.14) Ba	1.5 (0.2)	1.7 (0.2)	0.65 (0.05)	0.62 (0.06)
	5-25	6.9 (0.1) a	6.8 (0.2)	11.0 (1.2) a	10.7 (0.5) b	0.69 (0.59)	0.10 (0.03) ab	1.16 (0.11) ab	1.14 (0.04) b	1.6 (0.2)	1.8 (0.2)	0.66 (0.05)	0.70 (0.07)
	25-40	7.1 (0.1) Ab	6.9 (0.1) B	8.1 (1.1) Ab	6.2 (0.7) Bc	1.33 (1.20)	0.05 (0.02) b	0.87 (0.11) b	0.69 (0.06) c	1.6 (0.3)	1.7 (0.2)	0.68 (0.06)	0.74 (0.07)

3   Long-term tillage effects on the distribution of P fractions in German loess soils

3.3.2   Crop yields, labile P and CAL-P status of the sites

The wheat yields at Friemar, Lüttewitz and Zschortau (averaged from 2004 to 2010) were up to 16 % higher compared to the mean yield of Germany (7.6 Mg ha^{-1} winter wheat, *Statistisches Jahrbuch*, 2010), whereas the yields at Grombach were slightly lower (Table 5). The sugar beet yields under CT at all sites were higher compared to the mean yield of Germany (61.9 Mg ha^{-1} sugar beet, *Statistisches Jahrbuch*, 2011). Tillage had a significant effect on sugar beet yields (11% higher yields under CT compared to NT, $p \leq 0.05$), and the yields under NT at Grombach, Lüttewitz and Zschortau were less than the mean yield of Germany (Table 5).

The P status in 0-25 cm for all sites ranged from 15 to 58 mg CAL-P kg^{-1} soil (Table 5). According to *VDLUFA* (Association of German Agricultural Analytic and Research Institutes, 1997), the soils of three sites studied were in classes B (21-42 mg P kg^{-1}, Friemar (CT treatment) and Lüttewitz (CT and NT)) and C (45-58 mg P kg^{-1}, Zschortau (CT and NT)) under both treatments, suggesting that P supply at the sites was low to sufficient. At Friemar with 20 (NT) and Grombach, with 15 (CT) and 17 mg CAL-P kg^{-1} (NT), the required P amount for the crops is not directly provided (class A, ≤ 20 mg P kg^{-1}) according to *VDLUFA*. However, the *VDLUFA* recommendation was challenged by *Köster & Nieder* (2007). They calculated input and output budgets for a number of German arable soils from 1950 to 2003 and suggested that a P content of 30 to 50 mg kg^{-1} (class C in their system) may already assure a sufficient P supply. According to their system, the available P contents in soils at all sites ranged in the CAL-P content classes B and C. Additional data (P nutritional status of plants, yield responses to different P fertilization levels) would be required for a support of either the VDLUFA system or the system by *Köster & Nieder* (2007).

The depth distribution of CAL-P with higher P contents in the topsoil under NT compared to CT (Table 5) was in line with *Deubel* et al. (2011) and *Zorn* et al. (2011).

The increased P contents in the surface soil and thus reduced P contents in the subsurface soil of NT systems may result in reduced P availabilities for crops, especially in periods, where the surface soil is dried out and P side-root dressing may be required (*Zorn* et al., 2011).

The CAL-P content was negatively correlated with the Al$_\alpha$ content (r = -0.53), confirming that CAL-P is not bound to amorphous Al. The positive trend of the CAL-P content with the C$_{org}$ content (r = 0.46) suggests that the CAL method extracts also some P associated with organic matter. This was also true for the labile P fractions DI-P and NaHCO$_3$-P$_i$, which were also correlated with the total N content (r = 0.54 and r = 0.51 for DI-P and NaHCO$_3$-P$_i$, respectively). However, one has to keep in mind that correlation analyses do not necessarily reflect causal relationships. Similar depth gradients of the different P fractions and C$_{org}$ in the NT treatments due to residue and fertilizer incorporation in 0-5 cm may have also contributed to the correlation.

3.3.3 Effect of tillage and sampling depth on total P

Across both tillage systems, the weighted mean of the P$_t$ content was 667 mg kg^{-1} dry soil (Table 7). In the top 5 cm of the profile, the P$_t$ content was 15% higher under NT compared to CT, while in 5-25 cm and 25-40 cm, the differences of P$_t$ content between CT and NT were much smaller. The increased P$_t$ content under NT in the top 5 cm of the profile was associated with a significant decrease in P$_t$ with depth under NT, while under CT the differences between layers were much less pronounced. The main reason for the vertical stratification of P$_t$ contents under NT is most likely the lack of mixing by tillage. However, tillage-dependent differences in yield (Table 5) and thus less P removal by plants under NT may have also contributed to the differences in P stocks in the different depth ranges. A vertical stratification of P$_t$ in NT soils has been reported for soils of different textures and types (*O'Halloran*, 1993; *Redel* et al., 2007) and was suggested to be the result of the root distribution, the surface application of shoot biomass and fertilizers and the lack of mixing by tillage.

The gradients along the depth of labile P, stable P and residual P contents (Figure 3) and stocks were reflected by the gradient of P_t stocks with the smallest stocks in the topsoil and highest in the depth of 5 to 25 cm (which is at least partly due to the different thickness of the layers) (Table 7). Several studies reported higher P_t stocks under NT than CT (*Essington & Howard*, 2000 (0-60 cm); *Vu* et al., 2009 (0-40 cm)), which investigated mainly loess soils. Others did their research on loamy soils and did not find an effect (*Daroub* et al., 2001; *Muukonen* et al., 2007; *Shafquat & Pierzynski*, 2010).

The minimal effects of tillage on P_t contents and stocks in our study indicate that for sites with small erosion losses, the tillage system does not have marked effects on the P dynamics and distribution. In our study, potential soil losses by water erosion (*Schwertmann* et al., 1987) were estimated to range from 0.1 (NT) to 0.4 t (ha a)$^{-1}$ (CT) at Zschortau (the site with the smallest slope (0.25%) and annual precipitation (512 mm)), from 0.8 (NT) to 1.2 (ha a)$^{-1}$ (CT) at Friemar and from 1.4 (NT) to 3.5 (ha a)$^{-1}$ (CT) at Lüttewitz. The only site with a slightly increased potential erosion loss was Grombach with estimated losses of 4.7 (NT) and 8.4 (CT) t (ha a)$^{-1}$ (slope: 5.5%, annual precipitation: 776 mm). Using the P_t contents in the topsoil at the four sites, this corresponds to annual P losses ranging from 0.06 to 4.4 kg ha^{-1} under CT and 0.2 to 6.4 kg ha^{-1} under NT. However, only for the labile P content at Grombach under NT and CT, a gradient along the hillside length from low (about 50 mg kg^{-1} dry soil at the top) to high contents (about 150 mg kg^{-1} dry soil at the bottom) was detected (data not shown). The residual-P content was distributed in reversed order relating to hillside length (about 400 and 250 mg kg^{-1} at the top and the bottom, respectively). At a site (fine-loamy soils) with greater precipitation and a slope of 2%, marked differences between CT and NT

3 Long-term tillage effects on the distribution of P fractions in German loess soils

systems were reported (*Franklin* et al., 2007). The small erosion potential at our sites was affirmed by the low C_{org} contents in 25-40 cm at the lowest sampling point in the trial fields (data not shown), which indicated that no dislocation of soil inside the field had occurred at all four sites.

Table 7: Means, standard deviations and significance values of P fractions and P stocks (inorganic P: P_i, organic P: P_o) of both tillage treatments (conventional tillage: CT, no-till: NT) and depths (n = 4). Values followed by the same letter are not significantly different (p ≤0.05). Capital letters refer to the comparison of tillage treatments within one depth, while lower case letters refer to sampling depth within one tillage treatment.

P fraction	Tillage treatment	Contents (mg P kg⁻¹)			Stocks (kg ha⁻¹)
		0-5 cm	5-25 cm	25-40 cm	0-40 cm
Water extractable P (DI-P)	CT	9.2 (1.1) [A]	10.2 (0.7)	7.4 (1.9)	46.8 (6.0)
	NT	15.8 (1.7) [Ba]	8.6 (1.1) [b]	5.6 (1.7) [b]	45.8 (6.7)
Bicarbonate soluble P_i ($NaHCO_3$-P_i)	CT	36.5 (4.8)	34.7 (4.2)	29.3 (4.9)	171 (22)
	NT	48.3 (5.0) [a]	35.7 (2.4) [ab]	23.2 (4.5) [b]	181 (17)
Bicarbonate soluble P_o ($NaHCO_3$-P_o)	CT	55.8 (7.9)	71.3 (10.7)	60.2 (8.9)	341 (22)
	NT	64.2 (8.6)	66.6 (9.0)	56.5 (7.8)	352 (43)
Hydroxide soluble P_i (NaOH-P_i)	CT	93.4 (10.3)	91.9 (10.1)	79.1 (10.3)	454 (50)
	NT	114 (21)	100 (13)	77.2 (10.2)	523 (73)
Hydroxide soluble P_o (NaOH-P_o)	CT	93.4 (10.2) [A]	97.9 (6.3)	92.0 (9.9)	496 (32)
	NT	133 (2) [Ba]	105 (8) [a]	64.5 (13.3) [b]	517 (27)
Hydrochloric acid soluble P (HCl-P)	CT	152 (19)	158 (17)	141 (25)	787 (108)
	NT	161 (18)	153 (21)	137 (35)	837 (155)
Residual-P	CT	250 (28)	237 (28)	213 (28)	1198 (141)
	NT	256 (28)	229 (31)	210 (32)	1262 (169)
P_i	CT	541 (53)	532 (45)	470 (56)	2656 (260)
	NT	595 (65)	527 (49)	453 (49)	2849 (298)
P_o	CT	149 (17)	169 (15)	152 (4)	837 (51)
	NT	197 (7) [a]	172 (14) [ab]	121 (21) [b]	869 (68)
P_t	CT	691 (59)	701 (50)	622 (58)	3493 (275)
	NT	792 (68)	698 (45)	574 (68)	3719 (343)

3.3.4 Effect of tillage and sampling depth on labile, stable and residual-P

3.3.4.1 Labile P

The labile P content was highest in the upper 5 cm of the soils under NT compared to CT and decreased with sampling depth for both treatments, whereas below the top 5 cm, the differences between tillage treatments were minimal (Figure 5). This was also reported by *Zamuner* et al. (2008) for a clay loam and the most labile P fraction, extracted with an anion exchange membrane, was closely related to C_{org} contents. In our study, there was a strong correlation between the labile fraction DI-P and C_{org} content ($r = 0.55$) or total N content ($r = 0.54$), and between the labile fraction NaHCO$_3$-P$_i$ and total N content ($r = 0.51$). Moreover, a significant proportion of the variability (52%) of the labile P fractions (DI-P and NaHCO$_3$-P$_i$) was explained by C_{org} (Table 8) by models selected by stepwise multiple regression. However, the residue and P-fertilizer incorporation in 0-5 cm in the NT treatments may have also contributed to the correlations.

A comparison of the depth distributions of the P contents in the different fractions for the single sites and tillage treatments showed marked similarities for CAL-P, DI-P and NaHCO$_3$-P$_i$ (Table 9). In the NT treatments, P contents in all three fractions decreased considerably with depth at Friemar, Grombach and Lüttewitz, whereas at Zschortau, the contents in the fractions did either not consistently decrease with depth (CAL-P, NaHCO$_3$-P$_i$) or decreased only slightly (DI-P, Table 9). Under CT, changes of P contents in the three fractions with depth were generally much less pronounced, suggesting that all three fractions mobilize easily extractable P fractions. However, no simple arithmetic relationship was observed: the sum of DI-P and NaHCO$_3$-P$_i$ was generally higher than the CAL-P content (Table 9). Moreover, a suggestion that CAL-P extracts DI-P and parts of NaHCO$_3$-P$_i$ may only hold true for several sites and tillage systems (Table 9), but is not generally valid as indicated the data for Zschortau under NT, where CAL-P is higher than the sum of DI-P and NaHCO$_3$-P$_i$ in 0-5 and 5-25 cm (Table 9).

In contrast to the three fractions CAL-P, DI-P and NaHCO$_3$-P$_i$, NaHCO$_3$-P$_O$ (Table 9) and the other P fractions (data not shown) showed a different pattern: decreases of P contents with depth under NT did not occur or were much less pronounced than for the three fractions.

Table 8: Stepwise multiple regression analysis of P fractions (expressed in mg kg^{-1} soil) on soil properties, including clay content (clay; g kg^{-1} soil), oxalate-soluble Fe and Al (Fe$_{ox}$ and Al$_{ox}$; (mg kg^{-1} soil)), organic C (C$_{org}$ g kg^{-1} soil), carbonate-C (CO$_3$-C; g kg^{-1} soil) and pH (n = 72). Unit of intercepts is mg kg^{-1} dry soil.

		Equation				R^2
DI-P =	-5.28	- 0.012 clay p = 0.08	+ 0.0047 Fe$_{ox}$ p <0.0001	+ 0.90 C$_{org}$ p <0.0001		0.52
NaHCO$_3$-P$_i$ =	-13.8	+ 0.017 Fe$_{ox}$ p <0.0001	+ 1.89 C$_{org}$ p <0.0001			0.52
NaOH-P$_i$ =	249	- 29.8 pH p <0.0001	+ 0.18 clay p = 0.0003	- 0.057 Al$_{ox}$ p = 0.01	+ 0.020 Fe$_{ox}$ p = 0.002	0.54
HCl-P =	446	- 48.6 pH p <0.0001	+ 0.15 clay p = 0.01	+ 12.9 CO$_3$-C p <0.0001		0.47
Residual-P =	138	+ 0.73 clay p <0.0001	- 0.10 Al$_{ox}$ p = 0.07			0.45

The regressions of both organic P-fractions are not shown since the degrees of determination were very small (R^2 = 0.17 and 0.19 for NaHCO$_3$-P$_O$ and NaOH-P$_O$, respectively).

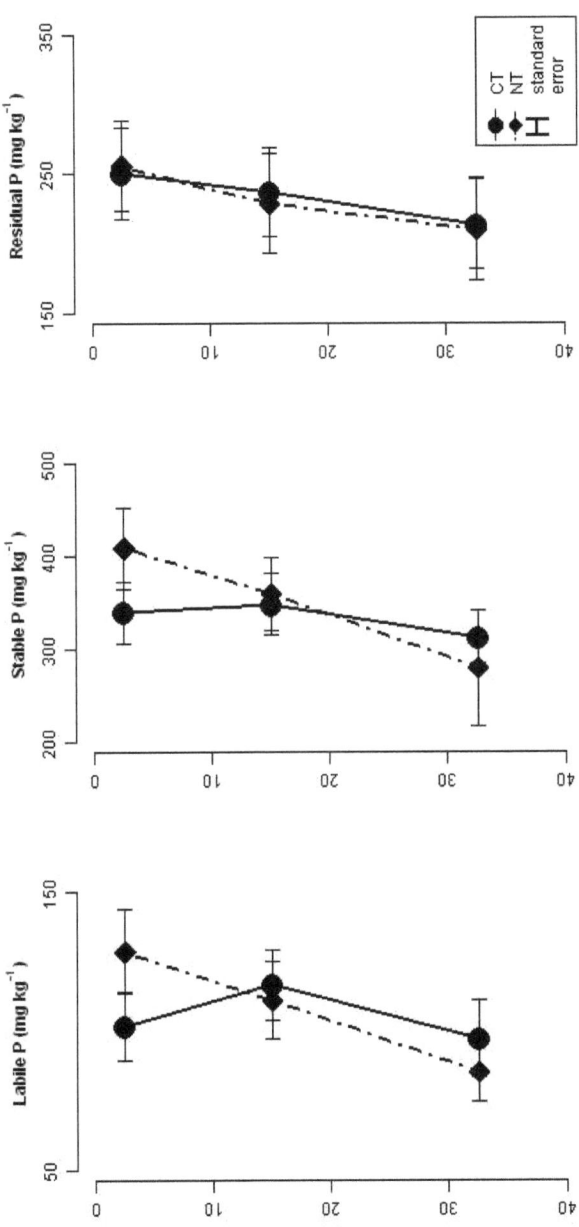

Figure 5: Labile (DI-P and NaHCO3-Pt), stable (NaOH-Pt and HCl-P) and residual-P content in the soil profile under conventional tillage (CT) and no-till (NT). Error bars refer to standard error (n = 4).

3   Long-term tillage effects on the distribution of P fractions in German loess soils

Table 9: Phosphorus contents in the fractions Calcium-Acetate-Lactate soluble P (CAL-P), water-soluble P (DI-P), and bicarbonate soluble inorganic and organic P (NaHCO$_3$-P$_i$ and NaHCO$_3$-P$_o$) for each site and tillage treatment (conventional tillage: CT; no-till: NT). Values shown are means and standard errors of the means of the pseudoreplicates (n = 3).

Site	Tillage treatment	Depth (cm)	CAL-P (mg kg^{-1})	DI-P (mg kg^{-1})	NaHCO$_3$-P$_i$ (mg kg^{-1})	NaHCO$_3$-P$_o$ (mg kg^{-1})
Friemar	CT	0-5	22.8 (4.9)	9.0 (1.6)	36 (8)	76 (31)
		5-25	20.4 (3.4)	9.6 (2.4)	33 (7)	100 (43)
		25-40	15.1 (2.7)	6.3 (1.7)	26 (8)	49 (19)
	NT	0-5	36.8 (6.6)	18.2 (4.6)	50 (13)	47 (15)
		5-25	15.3 (2.8)	6.0 (2.1)	29 (10)	40 (14)
		25-40	3.4 (0.2)	2.5 (0.4)	12 (4)	49 (28)
Grombach	CT	0-5	16.2 (2.7)	6.2 (1.9)	31 (10)	51 (17)
		5-25	14.1 (1.1)	8.9 (4.9)	31 (9)	50 (23)
		25-40	10.1 (1.2)	3.7 (2.0)	22 (7)	80 (54)
	NT	0-5	26.4 (1.0)	12.6 (2.8)	52 (4)	77 (24)
		5-25	14.9 (1.2)	7.8 (1.3)	37 (2)	77 (39)
		25-40	3.9 (1.3)	2.7 (0.9)	20 (3)	74 (35)
Lüttewitz	CT	0-5	44.7 (12.9)	11.3 (4.1)	50 (10)	57 (24)
		5-25	41.7 (13.3)	12.3 (3.6)	47 (10)	61 (24)
		25-40	36.6 (11.0)	12.8 (3.8)	44 (8)	69 (28)
	NT	0-5	65.8 (4.5)	19.3 (4.1)	57 (9)	81 (35)
		5-25	33.7 (8.8)	9.6 (2.7)	40 (7)	74 (37)
		25-40	22.9 (3.0)	8.5 (3.6)	32 (3)	64 (24)
Zschortau	CT	0-5	46.4 (5.4)	10.4 (3.4)	29 (5)	38 (10)
		5-25	44.9 (1.7)	10.1 (2.6)	29 (5)	74 (21)
		25-40	33.3 (2.4)	6.8 (1.2)	25 (4)	42 (13)
	NT	0-5	57.8 (3.4)	13.2 (1.4)	34 (4)	52 (20)
		5-25	57.9 (5.5)	10.8 (2.0)	37 (4)	76 (41)
		25-40	34.7 (7.0)	8.5 (3.4)	29 (6)	39 (17)

3.3.4.2 Stable P

The stable P contents were similarly distributed in the soil profile as the labile P contents with the highest contents in the upper 5 cm under NT compared to CT (Figure 3). However, the contents of stable P in CT and NT were three times higher than the contents of labile P. In contrast to labile P content, C_{org} was only weakly related to the stable P fractions. Even though C_{org} contributed significantly to NaOH-P$_o$ in the multivariate regression analysis, C_{org} was a poor predictor of NaOH-P$_o$ as the equation only explained 19% of the variability in NaOH-P$_o$. In contrast to our results, studies generally found that

C_{org} and P_o were correlated (*Tiessen* et al., 1984 (n = 168); *Burt* et al., 2002 (n = 21)). Our results suggest that the contribution of the organic fractions to P_o depended on tillage, depth, and/or site. A large proportion of P_o may have been associated with the light fraction, which consists of barely decomposed organic material (*Balesdent* et al., 2000).

The HCl-P content as one part of the stable P fraction amounted to 22.5% of P_t (averaged 0-40 cm) each for CT and NT. HCl-P is believed to consist mainly of Ca-associated P (*Hedley* et al., 1982 (black chernosemic soils); *Tiessen* et al., 1984; *Roberts* et al., 1985 (silty loam)). Our study supports this conclusion, as carbonates explained a significant proportion in the variability of HCl-P in the multivariate model (Table 8).

3.3.4.3 Residual-P

As expected, the distribution of residual-P in the soil profile was equal for CT and NT. The residual-P contents decreases slightly with soil depth. Our results are in line with *Wright* (2009) and *Vu* et al. (2009). Thus, tillage intensity had no effect on the most stable P fraction. The clay content was a significant factor for HCl-P and residual-P in the multivariate analysis (Table 8). The strong correlation between residual-P and clay content found in our study (r = 0.69) was also reported by *Tiessen* et al. (1984). Moreover, the Fe_{ox} and Al_{ox} contents did not explain a significant proportion of the variability of these two P fractions (Table 8), since oxalate extracts mainly amorphous P. In general, the stable P fractions, namely NaOH-P_i and HCl-P, as well as the most stable residual-P fraction, were mainly associated with soil constituents which were little affected by the tillage treatments. As these fractions accounted for two thirds of P_t, tillage could not have a large effect on P_t.

3 Long-term tillage effects on the distribution of P fractions in German loess soils

3.4 Conclusions

The tillage treatments had only little (and generally insignificant) effects on the total P content with a slightly increased soil P content under NT compared to CT. This was mainly due to an increase in labile P in the topsoil which may have resulted from a higher C_{org} content in the upper 5 cm. The absence of more pronounced effects was probably mainly because of the only small potential erosion losses at the sites studied. Therefore, for the climatic conditions and soil types prevalent at our study sites, factors other than P availability will determine the success of a tillage system.

Correlation analyses suggested that labile P contents were controlled by C_{org} while stable P and residual-P contents depended on the contents of Ca and clay, respectively.

3.5 References

AG Boden (Ed.) (2006): Bodenkundliche Kartieranleitung. 5. verbesserte und erweiterte Auflage. Schweizerbart'sche Verlagsbuchhandlung.

Balesdent, J., Chenu, C., Balabane, M. (2000): Relationship of soil organic matter dynamics to physical protection and tillage. *Soil & Tillage Research* 53, 215–230.

BMELV (Bundesministerium für Ernährung, L. u. V. (2010): Statistisches Jahrbuch über Ernährung, Landwirtschaft und Forsten der Bundesrepublik Deutschland. Wirtschaftsverlag NW GmbH, Bremerhaven (Germany).

BMELV (Bundesministerium für Ernährung, L. u. V. (2011): Statistisches Jahrbuch über Ernährung, Landwirtschaft und Forsten der Bundesrepublik Deutschland. Landwirtschaftsverlag GmbH Münster-Hiltrup, Münster (Germany).

Bowman, R. A., Cole, C. V. (1978): Transformations of Organic Phosphorus Substrates in Soils as Evaluated by NaHCO3 Extraktion. *Soil Science* 125, 49–54.

Bowman, R. A., Moir, J. O. (1993): Basic EDTA as an Extractant for Soil Organic Phosphorus. *Soil Science Society of America Journal* 57, 1516–1518.

Burt, R., Mays, M. D., Benham, E. C., Wilson, M. A. (2002): Phosphorus characterization and correlation with properties of selected benchmark soils of the United States. *Communications in Soil Science and Plant Aalysis* 33, 117–141.

Cross, A. F., Schlesinger, W. H. (1995): A literature-revies and evaluation of the Hedley fractionation - applications to the biogeochemical cycle of soil-phosphorus in natural ecosystems. *Geoderma* 64, 197–214.

D'Haene, K., Vermang, J., Cornelis, W. M., Leroy, B. L. M., Schiettecatte, W., Neve, S. de, Gabriels, D., Hofman, G. (2008): Reduced tillage effects on physical properties of silt loam soils growing root crops. *Soil & Tillage Research* 99, 279–290.

Daroub, S. H., Ellis, B. G., Robertson, G. P. (2001): Effect of cropping and low-chemical input systems on soil phosphorus fractions. *Soil Science* 166, 281–291.

Deubel, A., Hofmann, B., Orzessek, D. (2011): Long-term effects of tillage on stratification and plant availability of phosphate and potassium in a loess chernozem. *Soil & Tillage Research* 117, 85–92.

Dieckmann, J., Miller, H., Koch, H.-J. (2006): Rübenwachstum und Bodenstruktur. Ergebnisse aus dem Gemeinschaftsprojekt Bodenbearbeitung. *Zuckerindustrie* 131, 642–654.

DIN Deutsches Institut für Normung e.V. (1997): Bodenuntersuchungsverfahren im Landwirtschaftlichen Wasserbau - Chemische Laboruntersuchungen - Teil 6: Bestimmung des Gehaltes an oxalatlöslichem Eisen: DIN 19684-6:1997-12. Beuth Verlag, Berlin.

DIN Deutsches Institut für Normung e.V. (2001): Bodenbeschaffenheit - Bestimmung der Trockenrohdichte: DIN ISO 11272:2001-01. Beuth Verlag, Berlin.

Essington, M. E., Howard, D. D. (2000): Phosphorus availability and speciation in long-term no-till and disk-till soil. *Soil Science* 165, 144–152.

Franklin, D., Truman, C., Potter, T., Bosch, D., Strickland, T., Bednarz, C. (2007): Nitrogen and phosphorus runoff losses from variable and constant intensity rainfall simulations on loamy sand under conventional and strip tillage systems. *Journal of Environmental Quality* 36, 846–854.

Frizano, J., Johnson, A. H., Vann, Scatena, F. N. (2002): Soil phosphorus fractionation during forest development on landslide scars in the Luquillo Mountains, Puerto Rico. *Biotropica* 34, 17–26.

Guo, F. M., Yost, R. S. (1998): Partitioning soil phosphorus into three discrete pools of differing availability. *Soil Science* 163, 822–833.

Guo, F., Yost, R. S., Hue, N. V., Evensen, C. I., Silva, J. A. (2000): Changes in phosphorus fractions in soils under intensive plant growth. *Soil Science Society of America Journal* 64, 1681–1689.

Hedley, M. J., Stewart, J. W. B., Chauhan, B. S. (1982): Changes in inorganic and organic soil-phosphorus fractions induced by cultivation practices and by laboratory incubations. *Soil Science Society of America Journal* 46, 970–976.

Koch, H.-J., Dieckmann, J., Buechse, A., Maerlaender, B. (2009): Yield decrease in sugar beet caused by reduced tillage and direct drilling. *European Journal of Agronomy* 30, 101–109.

Köhn, M. (1928): Bemerkungen zur mechanischen Bodenanalyse. III. Ein neuer Pipettapparat. *Zeitschrift für Pflanzenernährung, Düngung, Bodenkunde* 11, 50–54.

Köster, W., Nieder, R. (2007): Wann ist eine Grunddüngung mit Phosphor, Kalium und Magnesium wirtschaftlich vertretbar? Hessisch-Oldendorf and Braunschweig (Germany). [cited 2011 Jun 21], Available from: http://www.beratung-mal-anders.de/pdf/Wann_ist_eine_Grundduengung_wirtschaftlich_DIN_A_4.pdf.

Mattingly, G. E. G. (1975): Labile Phosphate in Soils. *Soil Science* 119, 369–375.

Mengel, K., Kirkby, E. A. (2001): Principles of Plant Nutrition. Kluwer Academic Publishers, Bern, Switzerland.

Messiga, A. J., Ziadi, N., Angers, D. A., Morel, C., Parent, L.-E. (2011): Tillage practices of a clay loam soil affect soil aggregation and associated C and P concentrations. *Geoderma* 164, 225–231.

Murphy, J., Riley, J. P. (1962): A modified single solution method for the determination of phosphate in natural waters. *Analytica Chimica Acta* 27, 31–36.

Muukkonen, P., Hartikainen, H., Lahti, K., Sarkela, A., Puustinen, M., Alakukku, L. (2007): Influence of no-tillage on the distribution and lability of phosphorus in Finnish clay soils. *Agriculture Ecosystems & Environment* 120, 299–306.

O'Halloran, I. P. (1993): Effect of tillage and fertilization on inorganic and organic soil phosphorus. *Canadian Journal of Soil Science* 73, 359–369.

Redel, Y. D., Rubio, R., Rouanet, J. L., Borie, F. (2007): Phosphorus bioavailability affected by tillage and crop rotation on a Chilean volcanic derived Ultisol. *Geoderma* 139, 388–396.

Roberts, T. L., Stewart, J. W. B., Bettany, J. R. (1985): The Influence of Topography on the Distribution of Organic and Inorganic Soil Phosphorus Across a Narrow Environmental Gradient. *Canadian Journal of Soil Science* 65, 651–665.

SAS Institute (2010): SAS/STAT 9.22 User's Guide, Cary (NC). 8444 pp.

Sathya, S., Pitchai, G., James, I. R. (2009): Effect of soil properties on availability of nitrogen and phosphorus in submerged and upland soil – A review. *Agricultural Reviews* 30, 71–77.

Schmidt, J. P., Buol, S. W., Kamprath, E. J. (1996): Soil phosphorus dynamics during seventeen years of continuous cultivation: Fractionation analyses. *Soil Science Society of America Journal* 60, 1168–1172.

Scholz, G., Quinton, J. N., Strauss, P. (2008): Soil erosion from sugar beet in Central Europe in response to climate change induced seasonal precipitation variations. *Catena* 72, 91–105.

Schüller, H. (1969): Die CAL-Methode, eine neue Methode zur Bestimmung des pflanzenverfügbaren Phosphates in Böden. *Zeitschrift für Pflanzenernährung und Bodenkunde* 123, 48–63.

Schwertmann, U., Vogl W., Kainz, W. (1987): Bodenerosion durch Wasser. Vorhersage des Abtrags und Bewertung von Gegenmaßnahmen. Verlag Eugen Ulmer, Stuttgart (Germany).

Shafqat, M. N., Pierzynski, G. M. (2010): Long-Term Effects of Tillage and Manure Applications on Soil Phosphorus Fractions. *Communications in Soil Science and Plant Analysis* 41, 1084–1097.

Srivastava, O. P., Pathak, A. N. (1971): Available phosphorus in relation to forms of phosphate fractions in Uttar Pradesh soils. *Geoderma* 5, 287–296.

Tiessen, H., Moir, J. O. (1993): Characterization of available P by sequential extraction. In: Soil sampling and methods of analysis. Lewis Publishers, Boca Raton.

Tiessen, H., Stewart, J. W. B., Cole, C. V. (1984): Pathways of phosphorus transformations in soils of differing pedogenesis. *Soil Science Society of America Journal* 48, 853–858.

VDLUFA: Standpunkt "Phosphordüngung nach Bodenuntersuchung und Pflanzenbedarf". *Kerschberger, M.; Hege, U.; Jungk, A.* (eds.). Verband Deutscher Landwirtschaftlicher Untersuchungs- und Forschungsanstalten, Darmstadt (Germany).

Vu, D. T., Tang, C., Armstrong, R. D. (2009): Tillage system affects phosphorus form and depth distribution in three contrasting Victorian soils. *Australian Journal of Soil Research* 47, 33–45.

Wright, A. L. (2009): Phosphorus sequestration in soil aggregates after long-term tillage and cropping. *Soil & Tillage Research* 103, 406–411.

Zamuner, E. C., Picone, L. I., Echeverria, H. E. (2008): Organic and inorganic phosphorus in Mollisol soil under different tillage practices. *Soil & Tillage Research* 99, 131–138.

Zorn, W., Schröter, H., Wagner, S. (2011): Ohne Pflug anders düngen, in DLG-Mitteilungen 2011.

4 Effects of different cover crops and residue location on soil C and N dynamics analyzed within an incubation experiment

Christiane Piegholdt[1], Rouven Andruschkewitsch[1], Deborah Linsler[1], Bernard Ludwig[1]

[1]Department of Environmental Chemistry, University of Kassel, Nordbahnhofstr. 1a, 37213 Witzenhausen, Germany

key words: winter cover crops, incorporation, surface-application, crop residues, C and N mineralization, decomposition rate, organic matter

Abstract

The integration of cover crops in a crop rotation may affect soil C and N fluxes and pools. However, less is known about effects of cover crop specie and incorporation depth of plant residues as driven by the tillage intensity on soil C and N dynamics. The objective of this study was to analyze the effects of the biomass input from yellow mustard, phacelia, and oil radish in relation to the location of plant residues as placed on the soil surface or incorporated into the soil matrix on fractions indicative for labile C and N within an incubation experiment. Soil was collected from a cultivated field in North Hesse, Germany, and filled in columns. Therein the three cover crops were sown. After a growth period of 63 days, crops were incorporated or surface-applied to soil and the columns were frozen at -10 °C for 7 days, followed by an incubation period at 10 °C for 80 days. Biomass growth decreased in the order (aboveground yields in t / ha): mustard (7.0) > phacelia (5.7) > oil radish (4.4). Cumulative CO_2 emissions (1.93-2.93 t CO_2-C ha^{-1}) in the subsequent incubations showed a high variability and were not significantly different between the treatments, neither for the surface-applied nor for the incorporated residues, indicating that residue location did not have a marked effect on the

4 Effects of different cover crops and residue location on soil C and N dynamics analyzed within an incubation experiment

C mineralization. For the control treatments, significantly lower cumulative CO_2 emissions (0.51 t CO_2-C ha^{-1}) were measured. Cumulative N_2O emissions (0.10 to 0.44 kg N_2O-N ha^{-1}) were not significantly different between the treatments. Cumulative mineralized N was significantly and markedly less for all cover crops incorporated into the soil compared to the fallow treatment indicating an efficient prevention of nitrate leaching losses by the cover crops in fall to spring. However, for the surface-applied residues, only the treatment with mustard residues gave significantly smaller cumulative mineralized N than the control. Overall, the experiments confirmed a marked potential of all studied cover crops in the temporary preservation of nutrients. However, the residue location affected C and N dynamics only slightly.

4.1 Introduction

The cultivation of cover crops has several benefits for agricultural soils. Cover crops suppress weeds and pests, they may improve the soil structure, and increase the soil biological activity (*Sächsische Landesanstalt für Landwirtschaft*, 2004).

Next to the increased retention of organic C (C_{org}) and total N (N_t) in soil by cover crop residues (*Sainju* et al., 2008), they are also well known to repress land degradation by erosion (*Hoormann* et al., 2009; *Langdale* et al., 1991; *Power & Biederbeck*, 1991). Their roots and the crop residues cause a coarse surface structure and a higher potential infiltration rate leading to smaller surface runoff (*Cardoso* et al., 2012; *Hughes-Games & Bertrand*, 1991; *Villamil* et al., 2006). These factors reduce, for example, the mobilization, transport, and deposition of soil through water erosion.

The C and N dynamics, however, are not only affected by the biomass production of the different cover crops and the resulting enhanced litter input, but also by the location of the residues after the dying off of the crops. The location and related decomposition processes of the plant residues are heavily influenced by the tillage

4 Effects of different cover crops and residue location on soil C and N dynamics analyzed within an incubation experiment

intensity. For instance, *Jacobs* et al. (2011) reported for an incubation experiment with maize residues lower relative C losses and maize-C mineralization rates after 28 days of incubation under simulated minimum tillage compared to conventional tillage. Additionally, field studies also indicated that a decrease of tillage intensity by conversion of intensive to conservation tillage systems retains C_{org} and N_t in soils (*Salinas-Garcia* et al., 1997; *Watts* et al., 2010) because of decreased decomposition rates of organic matter (OM) (*Kladivko*, 2001; *Zibilske* et al., 2002).

Objectives of this study were to analyze the biomass production of mustard, phacelia, and oil radish and the effects of simulated increased tillage intensity (i.e., crop residue placement) on soil C and N dynamics after dying off of the different cover crops.

4.2 Material and methods

4.2.1 Site and soil

The soil for this pot experiment was collected in September 2010 from a cultivated field in Kleinalmerode (North Hesse, Germany) with a crop rotation consisting of two times winter wheat (*Triticum aestivum* L.) followed by rape (*Brassica napus* L.). At the sampling time, winter wheat was harvested and the soil was prepared for further sowing of winter wheat by mulching with a cultivator down to 15 cm. Glyphosat ($C_3H_8PNO_5$) as broad-spectrum herbicide for weed control was spread out in August 2010. In April 2010, 200 kg N ha^{-1} y^{-1} as ammonium sulfate fertilizer was applied to the field.

A composite soil sample (about 1.5 t) consisting of 8 sampling points was taken from 0-25 cm depth with a spade. The soil was a Podzol-Cambisol with 16% sand, 53% silt and 31% clay (silt clay loam, (Tu3); *AG Boden*, 2005). Stocks of C_{org} and N_t were calculated for 0-25 cm soil depth and accounted for 93 t C_{org} ha^{-1} and 7.1 t N_t ha^{-1}.

4 Effects of different cover crops and residue location on soil C and N dynamics analyzed within an incubation experiment

4.2.2 Experimental design

The soil was sieved at 10 mm, pre-incubated for one week at ambient temperature and filled in acrylic glass columns (Figure 6; 7 kg field moist soil per column, 5486 g dry soil, respectively; 14.55 cm diameter, height 25 cm; this corresponded to a bulk density of 1.5 g cm^{-3}), then adjusted to 60% of water holding capacity with deionized (DI) water and pre-incubated in climate chambers for two weeks at 10 °C.

After the pre-incubation, four different treatments were carried out: three cover crops, namely yellow mustard (*Sinapis alba* L.), phacelia (*Phacelia tanacetifolia* BENTH.), and oil radish (*Raphanus sativis* L.) were sown each in 15 replicates. Further 15 columns were left without seeds to simulate a fallow field (control treatment). To obtain the same plant density that is commonly regarded as good agricultural practice in the field (*Verband der Landwirtschaftskammern*, 2012), the cover crops were sown in the following densities: yellow mustard with 5, phacelia with 9, and oil radish with 3 plants per column. The seeding density was calculated as follows:

$$seeds = (rate \times area) \times (1000 \times TSW^{-1}) \qquad (2)$$

where *seeds* is the number of seeds (piece) per column, *rate* is the seed rate of the cover crops (g cm^{-2}) recommended of the *Verband der Landwirtschaftskammern* (2012; 20, 10, and 20 kg ha^{-1} for mustard, phacelia, and oil radish, respectively), *TSW* is the thousand-seed weight (g), which represents the weight of 1000 dry seeds (g) of the respective cover crop.

The growth of the cover crops took place in climate chambers with simulated day lengths and temperature gradients according to the sampling site. In regular intervals, the columns were re-adjusted to 60% of water holding capacity by adding deionized (DI) water. Because of obvious N deficiency of the cover crops (yellow leaves, reduced growth; *Christin* et al., 2009; *Pompelli* et al., 2010), a nutrient solution with ammonium nitrate (NH$_4$NO$_3$) was added in the 8th and 10th week of the growth phase to warrant a sufficient N supply (overall, 0.0444 g N per column, respectively 27 kg N ha^{-1}).

The three cover crops used in our study are commonly used in German agriculture and are known to be undemanding with respect to soil properties (*Sächsische Landesanstalt für Landwirtschaft*, 2004). All of them consist of a tap root, whereas mustard and oil radish (tap root down to 80-150 cm) are also rich in lateral roots compared to phacelia (tap root down to 60-80 cm), which are known to have only a few lateral roots.

4.2.3 Experimental set-up

The experimental set up consisted of three subsequent experimental periods: a growth period, a freezing period and an incubation period (an overview about soil sampling and analyses is given in Figure 6). The details are as follows:

T_0: start of the experiment: the first analysis for C and N contents were made on the field moist soil after one week pre-incubation (T_0), where soil samples were taken from three of the prepared columns directly after filling with the pre-incubated soil.

Period between T_0 and T_1 (0-63 days), growth period: the second sampling time was after the growth of the cover crops (T_1). The plants of 3 columns per cover crop were cut and slightly mixed with the soil of the first 15 cm in the column (incorporated crop residues; simulated mulching), to other 3 replicates per cover crop, the respective cut crop residues were surface-applied (simulated no-till). The control was treated as described for cover crop including columns, i.e. the soil of 0-15 cm was slightly mixed (comparable to incorporated crop residues) or left untreated (comparable to surface-applied crop residues).

Period between T_1 and T_2 includes freezing for 7 days (i.e., simulation of winter), followed by incubation at 10 °C (simulation of spring) for 80 days: after freezing of the soils at -10 °C for 7 days, the columns were left in the climate chamber for thawing to reach 4 °C for 1 day before incubation. Six replicates per treatment (3 cover crops and control) were handled as described for T_1 (i.e., 3 columns per cover crop and control for residue incorporation; 3 columns per cover crop and control for surface application of residues) and incubated at 10 °C for 80 days in climate chambers. Each column was placed on a ceramic plate (1-µm pore diameter) to which a constant suction of 100 hPa was applied. During the incubation period, each column was irrigated with 0.01 M $CaCl_2$ (2 mm d^{-1}). A constant suction (100 hPa) was applied to each column with flexible tubes and a pump, the irrigation solution was collected in glass bottles after leaching through the soil columns and frozen in polyethylene bottles until measurements. Percolates were collected weekly. After incubation, soil samples were taken from all columns (T_2).

4   Effects of different cover crops and residue location on soil C and N dynamics
analyzed within an incubation experiment

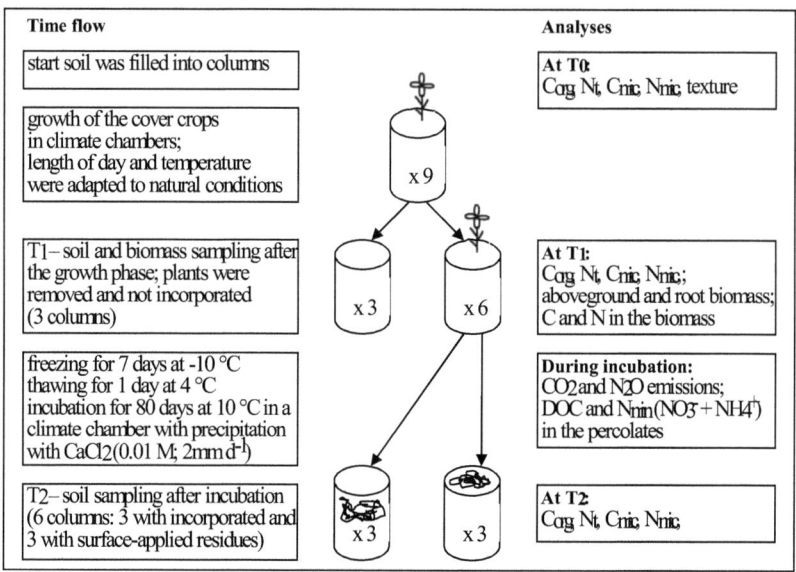

Figure 6: Scheme of the experimental design, sampling times and analyses carried out respectively for the three cover crop and the fallow treatment.

4.2.4 Measurement of aboveground and belowground biomass

Aboveground biomass was measured after the growth of the cover crops. The plants were harvested at T_1, dried at 40 °C and weighed. Belowground biomass was determined by hand sorting the roots from the soil at T_1. Therefore, the soil of each column at T_1 was vertically divided into two equal portions, both of them were weighed. One portion was sieved at 2 mm and stored at 4 °C until soil analyses. The second portion was carefully washed to separate the roots from the soil. The root-water suspension was filtered through 0.45 µm and the retained roots were dried at 40 °C and weighed. The C and N content of the aboveground and belowground biomass was determined by dry combustion (Elementar Vario El, Heraeus, Hanau, Germany).

4.2.5 Soil analyses

All soil analyses were conducted for soil sieved at 2 mm. A portion of the sampled field moist soil was dried at 40 °C and analyzed for texture following the pipet method (*DIN ISO 11277*, 2002). Field moist soil samples were analyzed for pH by extraction with $CaCl_2$ (20 g soil/ 50 ml 0.01 M $CaCl_2$). Gravimetric soil moisture content was determined by drying samples at 105 °C for 24 h. Bulk densities were determined according to *DIN ISO 11272* (2002).

Dry samples were used to determine total C and N content by dry combustion (Elementar Vario El, Heraeus, Hanau, Germany). Carbonate-C (CO_3-C) in soil was determined to calculate the C_{org} content as the difference between total C and CO_3-C. For CO_3-C determination, we followed the approach of *Walthert* et al. (2010) with slight modifications. Briefly, two 1 g-soil portions of dry ground soil were taken from each soil sample and weighed into a scintillation vial (20 ml). To one sub-sample, 2 ml of 1% HCl was added. Then, the open vials, as well as a glass beaker containing 100 ml of 32% HCl were placed into a desiccator, to expose the soil to the acid fumes for 24 h. After exposing, the glass beaker containing HCl was removed and the samples were dried for 24 h in a continuously vacuumized desiccator containing drying granules. Afterwards, the samples were completely dried at 40 °C for 3 days. Then, soil samples were washed with 100 ml of DI water to remove HCl, filtered through a polyamide filter (0.45 µm) and dried at 40 °C. A second sub-sample was directly washed with 100 ml of DI water without previous acid treatment and filtered as described for the first sub-sample. To obtain the complete C_{org} content of the soil samples, we determined dissolved organic C (DOC) in the filtrates of the washing step with a C/N analyzer (analytikjena multi N/C 2100S, Jena, Germany). The C content in the soil samples before and after HCl treatment was determined by dry combustion. CO_3-C contents were calculated as the difference between untreated soil samples and soil samples treated with HCl by also considering the contents of DOC in the filtrates.

4   Effects of different cover crops and residue location on soil C and N dynamics
analyzed within an incubation experiment

Microbial biomass C (C_{mic}) and N (N_{mic}) was determined by chloroform fumigation extraction (*Vance* et al., 1987). Briefly, two 5 g-soil portions were taken from each soil sample. One of them was directly extracted with 20 ml of 0.5 M K_2SO_4, while the other sub-sample was extracted after fumigation with $CHCl_3$ for 25 h at 25 °C. After filtration through folded filter paper (Whatman No. 595 ½), the extracts were analyzed for C and N with a C/N analyzer (analytikjena multi N/C 2100S, Jena, Germany). Microbial biomass C and N was calculated using the conversion factors of 0.45 for C_{mic} (*Joergensen*, 1996) and 0.54 for N_{mic} (*Brookes* et al., 1985), as the difference between fumigated and unfumigated samples.

4.2.6   *Measurement of CO_2 and N_2O emission, dissolved organic C and mineralized N*

During the incubation at 10 °C, emissions of CO_2 and N_2O were determined. Every 3.5 h, a gas sample was automatically taken with a P64 system (Loftfields Analytische Lösungen, Neu Eichenberg, Germany) and analyzed with a gas chromatograph (Shimadzu Gas Chromatograph GC-14A, Duisburg, Germany; flow: 10 ml min^{-1}) (*Sänger* et al., 2011).

Dissolved organic carbon (DOC) was measured in the percolates with a C/N analyzer (analytikjena multi N/C 2100S, Jena, Germany). Mineralized N (N_{min}) as the sum of nitrate (NO_3^-) and ammonia (NH_4^+) was measured in the percolates with a continuous flow analyzer (Evolution II auto-analyzer, Alliance Instruments, Cergy-Pontoise, France).

4.2.7   *Statistical analyses*

The data were analyzed with GNU R (*R Development Core Team*, 2010, Version 2.15.2) by Shapiro-Wilk normality test, analysis of variance (ANOVA) and correlation analysis, as a randomized design with tillage treatment as the main factor and cover crop species as

4 Effects of different cover crops and residue location on soil C and N dynamics analyzed within an incubation experiment

sub-factor. Some data sets were not normally distributed, so we conducted a logarithmic data transformation (boxcox transformation) to provide the preconditions (normal distribution and homogeneity of variance of the data sets) for a two-way ANOVA. Analysis of variance was performed on the average values of the three column replicates. For correlation analyses (Spearman's rank correlation), the three columns per cover crop and tillage treatment were used separately, to detect relationships of mineralized C and N to microbial biomass, C and N input, and cover crop yields. Effects were considered to be significant at $p \leq 0.05$.

Modeling of C mineralization to estimate the size of the labile C pools was conducted with a one-pool model using GNU R (*R Development Core Team*, 2010, Version 2.15.2). For the estimation of the decay constants we used a non-linear least square (nls) first-order model:

$$Y_{min}(t) = Y_l \times (1-e^{-k \times t}) \qquad (3)$$

where $Y_{min}(t)$ is C mineralized (kg ha^{-1}) at time t (days), Y_l is the labile C pool (kg ha^{-1}), k is the decay constant (day^{-1}). We followed the recommendation of *Wang* et al. (2003) to provide an unequivocal measure of soil C mineralization capacity, and fitted Eq (3) to the obtained data set of all cover crops and tillage simulations. We fixed the decay constants as the average of the single decay constants per cover crop treatment (n = 6) to the obtained values. The obtained decay constants under the cover crop treatments ranged between $k = 0.0239$ d^{-1} (for mustard and phacelia) and $k = 0.0264$ d^{-1} (for oil radish) for C mineralization.

4.3 Results and discussion

4.3.1 Biomass production of the cover crops, stocks of organic C (C_{org}) and total N (N_t), and stocks of microbial C and N (C_{mic}, N_{mic})

4 Effects of different cover crops and residue location on soil C and N dynamics analyzed within an incubation experiment

Aboveground biomass was highest for mustard (7.1 t ha^{-1}) followed by phacelia (5.7 t ha^{-1}) and oil radish (4.4 t ha^{-1}, Table 10). With regard to the differences in the amounts of aboveground and belowground biomass and the related variations in C and N input between the cover crop treatments (Table 10), differences in soil C and N dynamics in the subsequent incubation experiment after freezing were expected.

Table 10: Stocks and C/N ratios, as well as C and N input of aboveground and belowground biomass for soil samples of 0-25 cm depth. The data shown are mean values of the three column replicates and the standard errors of the means are given in parentheses (n = 3). Values followed by different letters are significantly different (p ≤0.05). Letters refer to the comparison of cover crop species.

Cover crop	Aboveground biomass				Root biomass			
	Stock	C-Input	N-Input	C/N ratio	Stock	C-Input	N-Input	C/N ratio
		(kg ha^{-1})				(kg ha^{-1})		
Mustard	7.0 (0.4)	2990 (180)	74.5 (8.3)	41 (3) a	4.3 (0.3)	960 (110)	40.7 (4.5)	24 (1)
Phacelia	5.7 (0.8)	2240 (340)	82.5 (7.6)	27 (2) b	4.5 (0.6)	1050 (110)	48.1 (5.4)	22 (1)
Oil Radish	4.4 (0.8)	1680 (290)	55.0 (6.0)	30 (2) b	5.2 (0.3)	1330 (90)	55.0 (5.6)	26 (1)

The C$_{mic}$ stocks ranged at T$_1$ from 1.92 to 2.08 t ha^{-1} (Table 11) and at T$_2$ from 0.94 to 2.31 t ha^{-1} (Table 12). At T$_2$, we found for the treatments with surface-applied residues significant higher C$_{mic}$ stocks in soils with phacelia (2.20 t ha^{-1}) compared to soils with mustard (0.94 t ha^{-1}). We found no significant differences between cover crops for the treatments with incorporated residues.

The N$_{mic}$ stocks increased under all cover crops and residue locations from the growth of the crops till the end of the incubation time, except in soils with surface-applied mustard residues. At T$_1$, N$_{mic}$ contents ranged from 0.37 to 0.42 t ha^{-1} (Table 11) and at T$_2$ from 0.50 to 0.66 t ha^{-1} (Table 12). We found no significant differences between cover crops or tillage treatments at T$_1$ and T$_2$.

Table 11: Stocks of organic C (C$_{org}$), total N (N$_t$), and microbial C and N (C$_{mic}$, N$_{mic}$) of soil samples from 0-25 cm depth of the three cover crop and the fallow after the growth of the crops (T_1). The data shown are mean values of the three column replicates and the standard errors are given in parentheses (n = 3).

Soil sampling	Cover crop	C_{org} (t ha^{-1})	N_t (t ha^{-1})	C_{mic} (t ha^{-1})	N_{mic} (t ha^{-1})
T0 (start)	-	93 (1)	7.1 (0)	2.07 (0.05)	0.49 (0.02)
T1 (after growth)	Fallow	95 (2)	7.3 (0.1)	1.92 (0.01)	0.37 (0)
	Mustard	93 (2)	7.1 (0.1)	1.86 (0.12)	0.40 (0.02)
	Phacelia	93 (2)	7.1 (0.1)	2.08 (0.02)	0.38 (0.01)
	Oil radish	91 (3)	7.1 (0.1)	2.01 (0.08)	0.42 (0.01)

4.3.2 Cumulative emissions of CO_2 and N_2O and release of DOC

The amounts of cumulative mineralized C showed a high variability and were unaffected by the residue location (Figure 7a and d). Cumulative CO_2-C emissions were well described by a one-pool model per cover crop treatment ($R^2 > 0.99$). Approximately 71% and 65% of the C added by the cover crops were mineralized in 80 days in the treatments with incorporated and surface-applied residues, respectively (potential priming effects were not considered). Similary, *Costa de Campos* et al. (2011) reported at slightly higher soil water contents under no-till compared to conventional tillage a comparable CO_2-C efflux from CT and NT soils. The significant lower cumulative amounts of mineralized C in the cover crop treatments compared to the fallow (under both residue locations) were also described by *Zhou* et al. (2012). We suggested higher concentrations of easily decopmposable C_{org} in the cover crops treatments compared with the fallow.

Cumulative N_2O-N emissions in the treatments with incorporated residues decreased in the order oil radish (0.37 kg ha^{-1}), mustard (0.18 kg ha^{-1}), fallow (0.16 kg ha^{-1}), and phacelia (0.10 kg ha^{-1}), however, the differences are lacking significance (Figure 7 b), which is similar to the differences observed for the treatments with surface-applied residues (Figure 7 e). Our results indicate only insignificant effects of the cover crops on N_2O emissions, independent from the residue location. Similarly,

4 Effects of different cover crops and residue location on soil C and N dynamics analyzed within an incubation experiment

Constantin et al. (2010) reported for three long-term experiments that cover crop establishment had no significant effect on gaseous N emissions. However, they found increased emissions for no-tillage treatments, which is generally only the case in poorly drained soils (*Rochette* et al. 2008; *Pelster* et al. 2011).

As expected, cumulative release of DOC was significantly less in the fallow treatment compared to the decomposition experiments with the cover crops (Table 12). The crop species, however, did not affect DOC releases significantly and variability was high (Table 12).

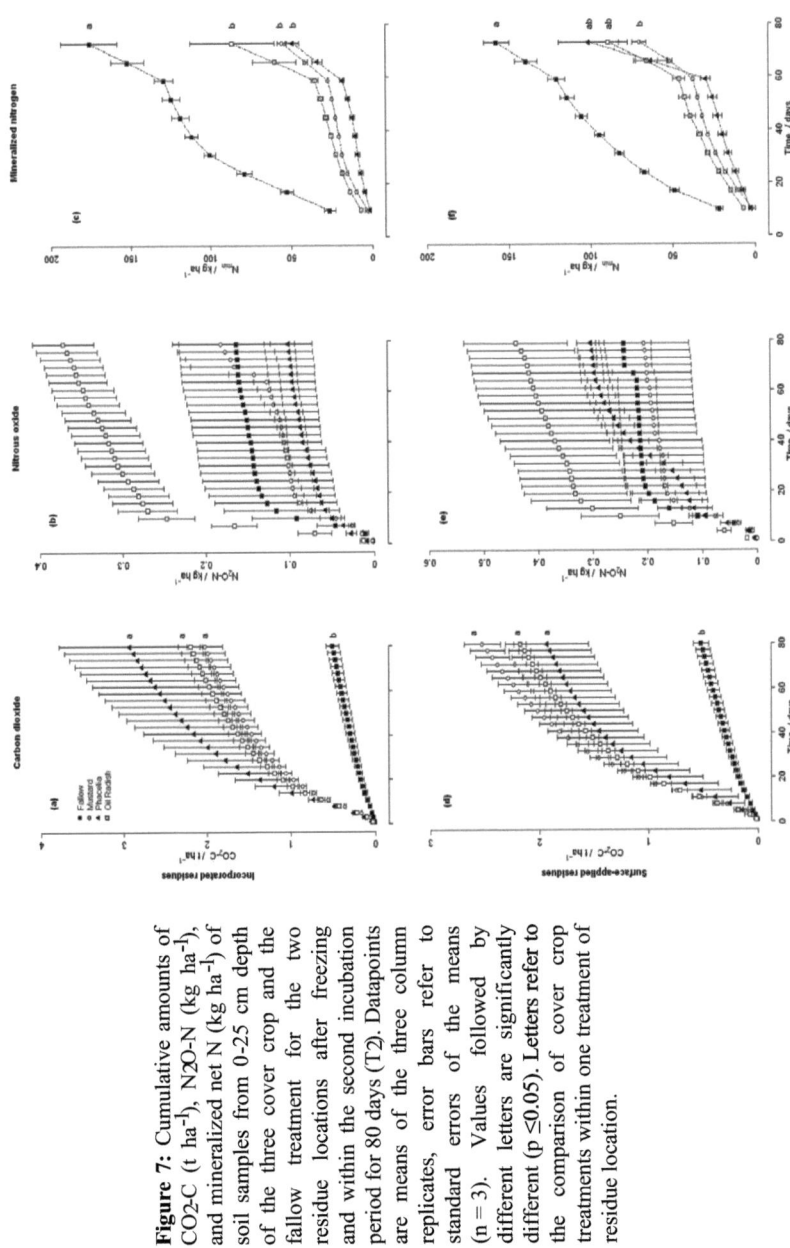

Figure 7: Cumulative amounts of CO_2-C (t ha^{-1}), N_2O-N (kg ha^{-1}), and mineralized net N (kg ha^{-1}) of soil samples from 0-25 cm depth of the three cover crop and the fallow treatment for the two residue locations after freezing and within the second incubation period for 80 days (T2). Datapoints are means of the three column replicates, error bars refer to standard errors of the means (n = 3). Values followed by different letters are significantly different (p ≤0.05). Letters refer to the comparison of cover crop treatments within one treatment of residue location.

4 Effects of different cover crops and residue location on soil C and N dynamics analyzed within an incubation experiment

4.3.3 Mineral N

Amounts of cumulative mineralized N were significantly highest in soils under fallow (176 kg N_{min} ha^{-1} compared to the soils with incorporated cover crop residues (49 to 86 kg ha^{-1}; Table 12; Figure 7 d and f), indicating the importance of cover crops for a reduction of N leaching (*Constantin* et al. 2010). This was in line with *Zhou* et al. (2012), who reported lower nitrate-N concentrations in the cover crop treatments compared to a fallow. However, for the surface-applied residues, only the treatment with mustard residues (70 kg N_{min} ha^{-1}) showed significantly smaller amounts of cumulative mineralized N than the control (157 kg ha^{-1}; Table 12).

Table 12: Stocks of leached dissolved organic carbon (DOC), mineralized N (N_{min}), and microbial C and N (C_{mic}, N_{mic}) of soil samples from 0-25 cm depth of the three cover crop and the fallow treatment for the two residue locations after freezing and within the second incubation period for 80 days (T2). The data shown are mean values of the three column replicates and the standard errors are given in parentheses (n = 3). Values followed by different letters are significantly different (p ≤0.05). Capital letters refer to the comparison of residue location within one cover crop treatment. Lower case letters refer to the comparison of cover crop treatments within one treatment of residue location.

Cover crop	Residue location	DOC (kg ha^{-1})	N_{min} (kg ha^{-1})	C_{mic} (t ha^{-1})	N_{mic} (t ha^{-1})
Fallow	control, mixed soil	13.6 (1.0)	176 (18) a	1.93 (0.06)	0.55 (0.05)
	control, unmixed soil	12.8 (0.6)	157(8) a	1.59 (0.41) ab	0.50 (0.12)
Mustard	incorporated	20.9 (7.5)	56 (3) b	2.24 (0.07) A	0.62 (0.05)
	surface-applied	17.6 (0.6)	70 (4) b	0.94 (0.19) Bb	0.33 (0.07)
Phacelia	incorporated	25.8 (1.7)	49 (4) Bb	2.31 (0.06)	0.63 (0.06)
	surface-applied	19.2 (1.5)	101 (18) Aab	2.20 (0.14) a	0.66 (0.05)
Oil radish	incorporated	19.3 (2.2)	86 (26) b	1.80 (0.26)	0.54 (0.08)
	surface-applied	20.6 (4.5)	89 (12) ab	1.79 (0.60) ab	0.60 (0.04)

4.3.4 Stocks of labile C

The C input in the different cover crop treatments depended on the biomass production in the growth period and inputs derived from roots and aboveground biomass (in t C ha^{-1}) decreased in the order mustard (4.0 t C ha^{-1}) > phacelia (3.3 t C ha^{-1}) > oil radish (3.0 t C

4 Effects of different cover crops and residue location on soil C and N dynamics analyzed within an incubation experiment

ha^{-1}, Table 10). Stocks of labile C were calculated from the C mineralization data by using a one-pool model and were in all incubation experiments with cover crops significantly higher than in the fallow soils, independent of the residue location (Figure 8). Also *Zhou* et al. (2012) modeled clearly higher labile organic C amounts in soils with cover crops compared to a control treatment without cover crops. Correlation analyses showed the labile C pool to be influenced by the input of C and N (r = 0.81, p <0.01 and r = 0.78, p <0.01, respectively).

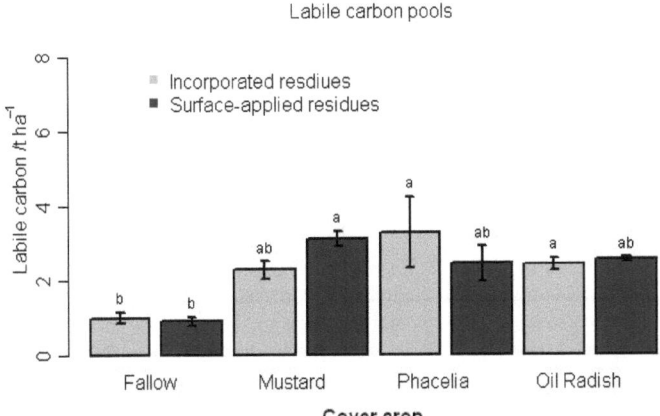

Figure 8: Stocks of labile C (based on cumulative CO_2-C emission) of soil samples from 0-25 cm depth of the three cover crop and the fallow treatment for the two residue locations after freezing and within the second incubation period for 80 days (T2). Columns show the mean values of the three column replicates, error bars refer to standard errors of the means (n = 3). Values followed by different letters are significantly different (p ≤0.05). Letters refer to the comparison of cover crop treatments within one treatment of residue location.

4 Effects of different cover crops and residue location on soil C and N dynamics analyzed within an incubation experiment

4.4 Conclusions

Our data indicate that the integration of winter cover crops in a crop rotation results in increased labile soil C and N stocks compared to crop rotations with a fallow period. The lower amounts of leached N_{min} in soils with cover crops compared to fallow soils indicated a strong benefit of cover crops on reducing nitrate-N losses by leaching between fall and spring. Additionally, the storage of mineralizable N in cover crops may lower the amounts of nitrogen fertilizers needed to match the N demand of the following crops.

4.5 References

AG Boden (Ed.) (2006): Bodenkundliche Kartieranleitung. 5. verbesserte und erweiterte Auflage. Schweizerbart'sche Verlagsbuchhandlung.

Brookes, P. C., Landman, A., Pruden, G., Jenkinson, D. S. (1985): Chloroform fumigation and the release of soil-nitrogen - a rapid direct extraction method to measure microbial biomass nitrogen in soil. *Soil Biology & Biochemistry* 17, 837–842.

Cardoso, D. P., Silva, M. L. N., Carvalho, G. J. d., Freitas, D. A. F. d., Avanzi, J. C. (2012): Cover crops to control soil, water and nutrient losses by water erosion. *Revista Brasileira de Engenharia Agricola e Ambiental* 16, 632–638.

Christin, H., Petty, P., Ouertani, K., Burgado, S., Lawrence, C., Kassem, M. A. (2012): Influence of Iron, Potassium, Magnesium, and Nitrogen Deficiencies on the Growth and Development of Sorghum (Sorghum bicolor L.) and Sunflower (Helianthus annuus L.) Seedlings. *Journal of Biotech Research*, 64–71.

Constantin, J., Mary, B., Laurent, F., Aubrion, G., Fontaine, A., Kerveillant, P., Beaudoin, N. (2010): Effects of catch crops, no till and reduced nitrogen fertilization on nitrogen leaching and balance in three long-term experiments. *Agriculture, Ecosystems & Environment* 135, 268–278.

Costa de Campos, B.-H., Amado, T. J. C., Tornquist, C. G., Nicoloso, R. d. S., Fiorin, J. E.: Long-term C-CO2 emissions and carbon crop residue mineralization in an oxisol under different tillage and crop rotation systems. *Revista Brasileira de Ciência do Solo* 35, 819–832.

DIN Deutsches Institut für Normung e.V. (2001): Bodenbeschaffenheit - Bestimmung der Trockenrohdichte: DIN ISO 11272:2001-01. Beuth Verlag, Berlin.

DIN Deutsches Institut für Normung e.V. (2002): Bodenbeschaffenheit - Bestimmung der Partikelgrößenverteilung in Mineralböden - Verfahren mittels Siebung und Sedimentation: DIN ISO 11277:2002-08. Beuth Verlag, Berlin.

Guadagnin, J. C., Bertol, I., Cassol, P. C., Amaral, A. J. d. (2005): Soil, water and nitrogen losses through erosion under different tillage systems. *Revista Brasileira de Ciência do Solo* 29, 277–286.

Hoormann, J. J., Islam, R., Sundermeier, A., Reeder, R. (2009): Using cover crops to convert to no-till. In: *The Ohio State University*: Fact sheet of Agriculture and natural Resources.

Hughes-Games, G. A., Bertrand, R. A. (1991): Effects of crop residue and tillage practices on water infiltration and crop production. In: *Hargrove, W. L.*: Cover crops for clean water. SWCS, Ankeny, IA, pp. 27–29.

Jacobs, A., Helfrich, M., Dyckmans, J., Rauber, R., Ludwig, B. (2011): Effects of residue location on soil organic matter turnover: results from an incubation experiment with N-15-maize. *Journal of Plant Nutrition and Soil Science* 174, 634–643.

Joergensen, R. G. (1996): The fumigation-extraction method to estimate soil microbial biomass: Calibration of the k(EC) value. *Soil Biology & Biochemistry* 28, 25–31.

Kladivko, E. J. (2001): Tillage systems and soil ecology. *Soil & Tillage Research* 61, 61–76.

Langdale, G. W., Blevins, R. L., Karlen, D. L., McCool, D. K., Nearing, M. A., Skidmore, E. L., Thomas, A. W., Tyler, D. D., Williams, J. R. (1991): Cover Crop effects on soil erosion by wind and water. In: *Hargrove, W. L.*: Cover crops for clean water. SWCS, Ankeny, IA, pp. 15–22.

Mills, W. C., Thomas, A. W., Langdale, G. W. (1991): Conservation tillage and season effects on soil-erosion risk. *Journal of Soil and Water Conservation* 46, 457–460.

Pelster, D. E., Larouche, F., Rochette, P., Chantigny, M. H., Allaire, S., Angers, D. A. (2011): Nitrogen fertilization but not soil tillage affects nitrous oxide emissions from a clay loam soil under a maize-soybean rotation. *Soil & Tillage Research* 115, 16–26.

Piotrowska, A., Wilczewski, E. (2012): Effects of catch crops cultivated for green manure and mineral nitrogen fertilization on soil enzyme activities and chemical properties. *Geoderma* 189, 72–80.

Pompelli, M. F., Martins, S. C. V., Antunes, W. C., Chaves, A. R. M., DaMatta, F. M. (2010): Photosynthesis and photoprotection in coffee leaves is affected by nitrogen and light availabilities in winter conditions. *Journal of Plant Physiology* 167, 1052–1060.

Power, J. F., Biederbeck, V. O. (1991): Role of cover crops in integrated crop production systems. In: *Hargrove, W. L.*: Cover crops for clean water. SWCS, Ankeny, IA, pp. 167–174.

R Development Core Team (2010): R: A language and environment for statistical computing.

Rochette, P., Angers, D. A., Chantigny, M. H., Bertrand, N. (2008): Nitrous oxide emissions respond differently to no-till in a loam and a heavy clay soil. *Soil Science Society of America Journal* 72, 1363–1369.

Saenger, A., Geisseler, D., Ludwig, B. (2011): Effects of moisture and temperature on greenhouse gas emissions and C and N leaching losses in soil treated with biogas slurry. *Biology and Fertility of Soils* 47, 249–259.

4 Effects of different cover crops and residue location on soil C and N dynamics analyzed within an incubation experiment

Sainju, U. M., Senwo, Z. N., Nyakatawa, E. Z., Tazisong, I. A., Reddy, K. C. (2008): Soil carbon and nitrogen sequestration as affected by long-term tillage, cropping systems, and nitrogen fertilizer sources. *Agriculture Ecosystems & Environment* 127, 234–240.

Sainju, U. M., Singh, B. P., Whitehead, W. F. (2002): Long-term effects of tillage, cover crops, and nitrogen fertilization on organic carbon and nitrogen concentrations in sandy loam soils in Georgia, USA. *Soil & Tillage Research* 63, 167–179.

SalinasGarcia, J. R., Hons, F. M., Matocha, J. E., Zuberer, D. A. (1997): Soil carbon and nitrogen dynamics as affected by long-term tillage and nitrogen fertilization. *Biology and Fertility of Soils* 25, 182–188.

Steele, M. K., Coale, F. J., Hill, R. L. (2012): Winter Annual Cover Crop Impacts on No-Till Soil Physical Properties and Organic Matter. *Soil Science Society of America Journal* 76, 2164–2173.

Vance, E. D., Brookes, P. C., Jenkinson, D. S. (1987): An Extraction Method for Measuring Soil Microbial Biomass-C. *Soil Biology & Biochemistry* 19, 703–707.

Verband der Landwirtschaftskammern (2012): Zwischenfrüchte für Futternutzung und Gründüngung. Hinweise zur artenwahl, Nutzungsmöglichkeiten und Anbauverfahren, http://www.lwk-niedersachsen.de/index.cfm/portal/2/nav/278/articale/21012.html.

Villamil, M. B., Bollero, G. A., Darmody, R. G., Simmons, F. W., Bullock, D. G. (2006): No-till corn/soybean systems including winter cover crops: Effects on soil properties. *Soil Science Society of America Journal* 70, 1936–1944.

Walthert, L., Graf, U., Kammer, A., Luster, J., Pezzotta, D., Zimmermann, S., Hagedorn, F. (2010): Determination of organic and inorganic carbon, delta C-13, and nitrogen in soils containing carbonates after acid fumigation with HCl. *Journal of Plant Nutrition and Soil Science* 173, 207–216.

Wang, W. J., Smith, C. J., Chen, D. (2003): Towards a standardised procedure for determining the potentially mineralisable nitrogen of soil. *Biology and Fertility of Soils* 37, 362–374.

Watts, D. B., Torbert, H. A., Prior, S. A., Huluka, G. (2010): Long-Term Tillage and Poultry Litter Impacts Soil Carbon and Nitrogen Mineralization and Fertility. *Soil Science Society of America Journal* 74, 1239–1247.

Zhou, X., Chen, C., Wu, H., Xu, Z. (2012): Dynamics of soil extractable carbon and nitrogen under different cover crop residues. *Journal of Soils and Sediments* 12, 844–853.

Zibilske, L. M., Bradford, J. M., Smart, J. R. (2002): Conservation tillage induced changes in organic carbon, total nitrogen and available phosphorus in a semi-arid alkaline subtropical soil. *Soil & Tillage Research* 66, 153–163.

5 General conclusions

The present thesis describe the influence of tillage intensity (i.e., conventional tillage: CT, reduced tillage: RT, and no-till: NT) on organic matter (OM) decomposition and stabilization, and on nutrient dynamic in arable loess soils. In general, significant differences in soil parameters between CT, RT, and NT were found in the surface soil (0-5 cm depth), which mainly depends on the tillage depth and crop residue distribution within the soil profile. According to the specific objectives (chapter 1.5), the main conclusions of the present thesis are:

(i) The analysis of soil samples from four long-term field experiment with 14 to 20 years of continuous tillage management and crop rotation and the fractionation of bulk soil OM leads to an improved comprehension of stability of the labile, intermediate, and passive OM pools in surface and subsurface soils. In surface soils, only NT resulted in larger stocks of the labile C and N pool compared to CT (chapter 2). The increase of labile pools in surface soils due to decreasing tillage intensity should have positive effects for the nutrient cycling and productivity, which is especially important for the development of more sustainable managed agro-ecosystems. These effects were not detected in subsurface soils indicating different effects of the tillage intensity on the OM dynamics with increasing soil depth. Most of C and N were located in the intermediate pool and accounted for 73 to 85% of the C stocks and for 70% to 95% of the N stocks. However, only for the surface soils, the stocks of the intermediate N pool were affected by the tillage intensity and distinctly larger for NT than for CT. The stocks of the passive C and N pool were not affected by the tillage intensity but were positively correlated to the stocks of the clay size fraction and oxalate soluble Al indicating a strong influence of site specific mineral characteristics on the size of these pools. Our results imply that the potential benefits of decreasing tillage intensity with respect to soil functions that are

5 General conclusions

closely related to OM dynamics have to be evaluated separately for surface and subsurface soils.

(ii) Next to the effects of tillage on C and N mineralization, also the P dynamic in soil is affected by the tillage system. Within 0-40 cm soil depth, the total P concentration in soil differ not between the tillage systems. For surface soils, the total P content was 15% higher for NT than for CT while with increasing depth the decrease in the total P content was stronger under NT than under CT. The higher P contents in the topsoil of NT presumably resulted from the shallower incorporation of harvest residues and fertilizer P compared to CT, whereas estimated soil losses and thus also P losses due to water erosion were only small for both treatments (chapter 3). Overall, only small differences in the P fractions and availability were observed between NT and CT. In contrast, contents of oxalate extractable iron and organic carbon were positively related to the labile fractions of inorganic P, while there was a high correlation of the stable inorganic P fractions with the clay contents and pH. This indicates a strong influence of site specific soil mineral characteristics on the labile and stable inorganic P fractions beside management sensitive parameters such as the organic carbon content and the pH. Because of the generally low influence of the tillage intensity on the P availability, other factors will determine the success of a tillage system under the given climatic and pedogenetic conditions.

(iii) The decomposition experiment provided information about the mineralization of biomass from different cover crop species (yellow mustard, phacelia, and oil radish) in relation to the location of plant residues (residues on the soil surface (comparable to NT) and incorporated into the soil (comparable to RT)). Cumulative CO_2 and N_2O emissions from the soils of the different treatments showed a high variability and were not significantly different between the treatments independent from the simulated tillage intensity. This indicates that the cover species and the residue location did not influence greenhouse gas emissions. Pronounced higher CO_2 emissions were found from soils with cover crops compared to fallow soils, assumedly caused by increased labile C and N stocks in soils including cover crops (chapter 4). Cumulative mineralized N was significantly and markedly less for all cover crops incorporated into the soil compared to the fallow treatment indicating an efficient prevention of nitrate leaching losses under RT. In contrast, for the surface-applied residues, only the treatment with mustard residues gave significantly smaller cumulative mineralized N than the fallow treatment. This indicates that the temporary preservation potential of nutrient is influenced by the tillage intensity derived location of catch crop residues.

Summarized, the present work showed the potential of reduced tillage systems to store carbon and nitrogen in the labile and intermediate organic matter pool, as well as plant available phosphorus, mainly in the soils.

References

AG Boden (Ed.) (2006): Bodenkundliche Kartieranleitung. 5. verbesserte und erweiterte Auflage. Schweizerbart'sche Verlagsbuchhandlung.

Alvarez, R., Santanatoglia, O. J., Daniel, P. E., Garcia, R. (1995): Respiration and specific activity of soil microbial biomass under conventional and reduced tillage. Pesquisa Agropecuaria Brasileira 30, 701–709.

Andruschkewitsch, R., Geisseler, D., Koch, H.-J., Ludwig, B. (2013): Effects of tillage on contents of organic carbon, nitrogen, water-stable aggregates and light fraction for four different long-term trials. Geoderma 192, 368–377.

Bailey, V. L., Smith, J. L., Bolton, H. (2002): Fungal-to-bacterial ratios in soils investigated for enhanced C sequestration. Soil Biology & Biochemistry 34, 997–1007.

Baker, J. M., Ochsner, T. E., Venterea, R. T., Griffis, T. J. (2007): Tillage and soil carbon sequestration - What do we really know? Agriculture Ecosystems & Environment 118, 1–5.

Balesdent, J., Chenu, C., Balabane, M. (2000): Relationship of soil organic matter dynamics to physical protection and tillage. Soil & Tillage Research 53, 215–230.

Balesdent, J., Pétraud J. P., Feller C. (1991): Effets des ultrasons sur la distribution granulométrique des matières organiques des sols. Science du Sol 29, 95–106.

Balota, E. L., Colozzi, A., Andrade, D. S., Dick, R. P. (2004): Long-term tillage and crop rotation effects on microbial biomass and C and N mineralization in a Brazilian Oxisol. Soil & Tillage Research 77, 137–145.

BMELV (Bundesministerium für Ernährung, L. u. V. (2010): Statistisches Jahrbuch über Ernährung, Landwirtschaft und Forsten der Bundesrepublik Deutschland. Wirtschaftsverlag NW GmbH, Bremerhaven (Germany).

BMELV (Bundesministerium für Ernährung, L. u. V. (2011): Statistisches Jahrbuch über Ernährung, Landwirtschaft und Forsten der Bundesrepublik Deutschland. Landwirtschaftsverlag GmbH Münster-Hiltrup, Münster (Germany).

Bowman, R. A., Cole, C. V. (1978): Transformations of Organic Phosphorus Substrates in Soils as Evaluated by NaHCO3 Extraktion. Soil Science 125, 49–54.

Bowman, R. A., Moir, J. O. (1993): Basic EDTA as an Extractant for Soil Organic Phosphorus. Soil Science Society of America Journal 57, 1516–1518.

References

Brookes, P. C., Landman, A., Pruden, G., Jenkinson, D. S. (1985): Chloroform fumigation and the release of soil-nitrogen - a rapid direct extraction method to measure microbial biomass nitrogen in soil. *Soil Biology & Biochemistry* 17, 837–842.

Brye, K.R., Longer, D.E. & Gbur, E.E. 2006. Impact of tillage and residue burning on carbon dioxide flux in a wheat-soybean production system. *Soil Science Society of America Journal*, **70**, 1145–1154.

Burt, R., Mays, M. D., Benham, E. C., Wilson, M. A. (2002): Phosphorus characterization and correlation with properties of selected benchmark soils of the United States. *Communications in Soil Science and Plant Aalysis* 33, 117–141.

Cambardella, A. C. A., Elliott, E. T. (1993): Carbon and nitrogen distribution in aggregates from cultivated and native grassland soils. *Soil Science Society of America Journal* 57, 1071–1076.

Cardoso, D. P., Silva, M. L. N., Carvalho, G. J. d., Freitas, D. A. F. d., Avanzi, J. C. (2012): Cover crops to control soil, water and nutrient losses by water erosion. *Revista Brasileira de Engenharia Agricola e Ambiental* 16, 632–638.

Christin, H., Petty, P., Ouertani, K., Burgado, S., Lawrence, C., Kassem, M. A. (2012): Influence of Iron, Potassium, Magnesium, and Nitrogen Deficiencies on the Growth and Development of Sorghum (Sorghum bicolor L.) and Sunflower (Helianthus annuus L.) Seedlings. *Journal of Biotech Research*, 64–71.

Constantin, J., Mary, B., Laurent, F., Aubrion, G., Fontaine, A., Kerveillant, P., Beaudoin, N. (2010): Effects of catch crops, no till and reduced nitrogen fertilization on nitrogen leaching and balance in three long-term experiments. *Agriculture, Ecosystems & Environment* 135, 268–278.

Coppens, F., Merckx, R., Recous, S. (2006): Impact of crop residue location on carbon and nitrogen distribution in soil and in water-stable aggregates. *European Journal of Soil Science* 57, 570–582.

Costa de Campos, B.-H., Amado, T. J. C., Tornquist, C. G., Nicoloso, R. d. S., Fiorin, J. E.: Long-term C-CO2 emissions and carbon crop residue mineralization in an oxisol under different tillage and crop rotation systems. *Revista Brasileira de Ciência do Solo* 35, 819–832.

Cross, A. F., Schlesinger, W. H. (1995): A literature-revies and evaluation of the Hedley fractionation - applications to the biogeochemical cycle of soil-phosphorus in natural ecosystems. *Geoderma* 64, 197–214.

References

Curtin, D., Wang, H., Selles, F., McConkey, B.G. & Campbell, C.A. 2000. Tillage effects on carbon fluxes in continuous wheat and fallow-wheat rotations. *Soil Science Society of America Journal*, **64**, 2080–2086.

D'Haene, K., Vermang, J., Cornelis, W. M., Leroy, B. L. M., Schiettecatte, W., Neve, S. de, Gabriels, D., Hofman, G. (2008): Reduced tillage effects on physical properties of silt loam soils growing root crops. *Soil & Tillage Research* 99, 279–290.

Daroub, S. H., Ellis, B. G., Robertson, G. P. (2001): Effect of cropping and low-chemical input systems on soil phosphorus fractions. *Soil Science* 166, 281–291.

Deubel, A., Hofmann, B., Orzessek, D. (2011): Long-term effects of tillage on stratification and plant availability of phosphate and potassium in a loess chernozem. *Soil & Tillage Research* 117, 85–92.

Dieckmann, J., Miller, H., Koch, H.-J. (2006): Rübenwachstum und Bodenstruktur. Ergebnisse aus dem Gemeinschaftsprojekt Bodenbearbeitung. *Zuckerindustrie* 131, 642–654.

DIN Deutsches Institut für Normung e.V. (1997): Bodenuntersuchungsverfahren im Landwirtschaftlichen Wasserbau - Chemische Laboruntersuchungen - Teil 6: Bestimmung des Gehaltes an oxalatlöslichem Eisen: DIN 19684-6:1997-12. Beuth Verlag, Berlin.

DIN Deutsches Institut für Normung e.V. (2001): Bodenbeschaffenheit - Bestimmung der Trockenrohdichte: DIN ISO 11272:2001-01. Beuth Verlag, Berlin.

DIN Deutsches Institut für Normung e.V. (2002): Bodenbeschaffenheit - Bestimmung der Partikelgrößenverteilung in Mineralböden - Verfahren mittels Siebung und Sedimentation: DIN ISO 11277:2002-08. Beuth Verlag, Berlin.

DIN Deutsches Institut für Normung e.V. (2009): Bodenbeschaffenheit - Felduntersuchungen - Teil 13: Bestimmung der Carbonate, der Sulfide, des pH-Wertes und der Eisen(II)-Ionen: DIN 19682-13:2009-01. Beuth Verlag, Berlin.

Ellert, B. H., Bettany, JR (1995): Calculation of organic matter and nutrients stored in soils under contrasting management regimes. *Canadian Journal of Soil Science* 75, 529–538.

Essington, M. E., Howard, D. D. (2000): Phosphorus availability and speciation in long-term no-till and disk-till soil. *Soil Science* 165, 144–152.

Fierer, N., Schimel, J. P. (2002): Effects of drying-rewetting frequency on soil carbon and nitrogen transformations. *Soil Biology & Biochemistry* 34, 777–787.

References

Franklin, D., Truman, C., Potter, T., Bosch, D., Strickland, T., Bednarz, C. (2007): Nitrogen and phosphorus runoff losses from variable and constant intensity rainfall simulations on loamy sand under conventional and strip tillage systems. *Journal of Environmental Quality* 36, 846–854.

Frizano, J., Johnson, A. H., Vann, Scatena, F. N. (2002): Soil phosphorus fractionation during forest development on landslide scars in the Luquillo Mountains, Puerto Rico. *Biotropica* 34, 17–26.

Green, V. S., Stott, D. E., Cruz, J. C., Curi, N. (2007): Tillage impacts on soil biological activity and aggregation in a Brazilian Cerrado Oxisol. *Soil & Tillage Research* 92, 114–121.

Guadagnin, J. C., Bertol, I., Cassol, P. C., Amaral, A. J. d. (2005): Soil, water and nitrogen losses through erosion under different tillage systems. *Revista Brasileira de Ciência do Solo* 29, 277–286.

Guo, F. M., Yost, R. S. (1998): Partitioning soil phosphorus into three discrete pools of differing availability. *Soil Science* 163, 822–833.

Guo, F., Yost, R. S., Hue, N. V., Evensen, C. I., Silva, J. A. (2000): Changes in phosphorus fractions in soils under intensive plant growth. *Soil Science Society of America Journal* 64, 1681–1689.

Hargrove, W. L. (Ed.) (1991): Cover crops for clean water. SWCS, Ankeny, IA.

Hedley, M. J., Stewart, J. W. B., Chauhan, B. S. (1982): Changes in inorganic and organic soil-phosphorus fractions induced by cultivation practices and by laboratory incubations. *Soil Science Society of America Journal* 46, 970–976.

Heitkamp, F., Raupp, J., Ludwig, B. (2009): Impact of fertilizer type and rate on carbon and nitrogen pools in a sandy Cambisol. *Plant and Soil* 319, 259–275.

Helfrich, M., Flessa, H., Mikutta, R., Dreves, A., Ludwig, B. (2007): Comparison of chemical fractionation methods for isolating stable soil organic carbon pools. *European Journal of Soil Science* 58, 1316–1329.

Hermle, S., Anken, T., Leifeld, J., Weisskopf, P. (2008): The effect of the tillage system on soil organic carbon content under moist, cold-temperate conditions. *Soil & Tillage Research* 98, 94–105.

Hoormann, J. J., Islam, R., Sundermeier, A., Reeder, R. (2009): Using cover crops to convert to no-till, in The Ohio State University: Fact sheet of Agriculture and natural Resources.

References

Hughes-Games, G. A., Bertrand, R. A. (1991): Effects of crop residue and tillage practices on water infiltration and crop production, in Hargrove, W. L.: Cover crops for clean water. SWCS, Ankeny, IA, pp. 27–29.

Jacobs, A., Helfrich, M., Hanisch, S., Quendt, U., Rauber, R., Ludwig, B. (2010): Effect of conventional and minimum tillage on physical and biochemical stabilization of soil organic matter. *Biology and Fertility of Soils* 46, 671–680.

Jacobs, A., Helfrich, M., Dyckmans, J., Rauber, R., Ludwig, B. (2011): Effects of residue location on soil organic matter turnover: results from an incubation experiment with N-15-maize. *Journal of Plant Nutrition and Soil Science* 174, 634–643.

Jagadamma, S., Lal, R. (2010): Integrating physical and chemical methods for isolating stable soil organic carbon. *Geoderma* 158, 322–330.

Janzen, H. H. (2004): Carbon cycling in earth systems - a soil science perspective. *Agriculture Ecosystems & Environment* 104, 399–417.

Joergensen, R. G. (1996): The fumigation-extraction method to estimate soil microbial biomass: Calibration of the k(EC) value. *Soil Biology & Biochemistry* 28, 25–31.

Kader, M. A., Sleutel, S., Begum, S. A., D'Haene, K., Jegajeevagan, K., Neve, S. de (2010): Soil organic matter fractionation as a tool for predicting nitrogen mineralization in silty arable soils. *Soil Use and Management* 26, 494–507.

Kaiser, K., Mikutta, R., Guggenberger, G. (2007): Increased stability of organic matter sorbed to ferrihydrite and goethite on aging. *Soil Science Society of America Journal* 71, 711–719.

Kaiser, M., Ellerbrock, R. H., Wulf, M., Dultz, S., Hierath, C., Sommer, M. (2012): The influence of mineral characteristics on organic matter content, composition, and stability of topsoils under long-term arable and forest land use. *Journal of Geophysical Research-Biogeosciences* 117.

Kaiser, M., Wirth, S., Ellerbrock, R. H., Sommer, M. (2010): Microbial respiration activities related to sequentially separated, particulate and water-soluble organic matter fractions from arable and forest topsoils. *Soil Biology & Biochemistry* 42, 418–428.

Kladivko, E. J. (2001): Tillage systems and soil ecology. *Soil & Tillage Research* 61, 61–76.

Kleber, M., Mikutta, R., Torn, M. S., Jahn, R. (2005): Poorly crystalline mineral phases protect organic matter in acid subsoil horizons. *European Journal of Soil Science* 56, 717–725.

References

Koch, H.-J., Dieckmann, J., Buechse, A., Maerlaender, B. (2009): Yield decrease in sugar beet caused by reduced tillage and direct drilling. *European Journal of Agronomy* 30, 101–109.

Koegel-Knabner, I., Guggenberger, G., Kleber, M., Kandeler, E., Kalbitz, K., Scheu, S., Eusterhues, K., Leinweber, P. (2008): Organo-mineral associations in temperate soils: Integrating biology, mineralogy, and organic matter chemistry. *Journal of Plant Nutrition and Soil Science* 171, 61–82.

Köhn, M. (1928): Bemerkungen zur mechanischen Bodenanalyse. III. Ein neuer Pipettapparat. *Zeitschrift für Pflanzenernährung, Düngung, Bodenkunde* 11, 50–54.

Köster, W., Nieder, R. (2007): Wann ist eine Grunddüngung mit Phosphor, Kalium und Magnesium wirtschaftlich vertretbar? Hessisch-Oldendorf and Braunschweig (Germany). [cited 2011 Jun 21], Available from: http://www.beratung-mal-anders.de/pdf/Wann_ist_eine_Grundduengung_wirtschaftlich_DIN_A_4.pdf.

Langdale, G. W., Blevins, R. L., Karlen, D. L., McCool, D. K., Nearing, M. A., Skidmore, E. L., Thomas, A. W., Tyler, D. D., Williams, J. R. (1991): Cover Crop effects on soil erosion by wind and water, in Hargrove, W. L.: Cover crops for clean water. SWCS, Ankeny, IA, pp. 15–22.

Luo, Z., Wang, E., Sun, O. J. (2010): Can no-tillage stimulate carbon sequestration in agricultural soils? A meta-analysis of paired experiments. *Agriculture Ecosystems & Environment* 139, 224–231.

Mattingly, G. E. G. (1975): Labile Phosphate in Soils. *Soil Science* 119, 369–375.

Messiga, A. J., Ziadi, N., Angers, D. A., Morel, C., Parent, L.-E. (2011): Tillage practices of a clay loam soil affect soil aggregation and associated C and P concentrations. *Geoderma* 164, 225–231.

Mikha, M. M., Rice, C. W. (2004): Tillage and manure effects on soil and aggregate-associated carbon and nitrogen. *Soil Science Society of America Journal* 68, 809–816.

Mikutta, R., Kleber, M., Torn, M. S., Jahn, R. (2006): Stabilization of soil organic matter: Association with minerals or chemical recalcitrance? *Biogeochemistry* 77, 25–56.

Mikutta, R., Zang, U., Chorover, J., Haumaier, L., Kalbitz, K. (2011): Stabilization of extracellular polymeric substances (Bacillus subtilis) by adsorption to and coprecipitation with Al forms. *Geochimica et Cosmochimica Acta* 75, 3135–3154.

Mills, W. C., Thomas, A. W., Langdale, G. W. (1991): Conservation tillage and season effects on soil-erosion risk. *Journal of Soil and Water Conservation* 46, 457–460.

References

Murphy, J., Riley, J. P. (1962): A modified single solution method for the determination of phosphate in natural waters. *Analytica Chimica Acta* 27, 31–36.

Muukkonen, P., Hartikainen, H., Lahti, K., Sarkela, A., Puustinen, M., Alakukku, L. (2007): Influence of no-tillage on the distribution and lability of phosphorus in Finnish clay soils. *Agriculture Ecosystems & Environment* 120, 299–306.

O.P. Srivastava, Pathak, A. N. (1971): Available phosphorus in relation to forms of phosphate fractions in Uttar Pradesh soils. *Geoderma* 5, 287–296.

O'Halloran, I. P. (1993): Effect of tillage and fertilization on inorganic and organic soil phosphorus. *Canadian Journal of Soil Science* 73, 359–369.

Oorts, K., Garnier, P., Findeling, A., Mary, B., Richard, G., Nicolardot, B. (2007): Modeling soil carbon and nitrogen dynamics in no-till and conventional tillage using PASTIS model. *Soil Science Society of America Journal* 71, 336–346.

Pandey, D., Agrawal, M. & Bohra, J.S. 2012. Greenhouse gas emissions from rice crop with different tillage permutations in rice-wheat system. *Agriculture Ecosystems & Environment* 159, 133–144.

Pelster, D. E., Larouche, F., Rochette, P., Chantigny, M. H., Allaire, S., Angers, D. A. (2011): Nitrogen fertilization but not soil tillage affects nitrous oxide emissions from a clay loam soil under a maize-soybean rotation. *Soil & Tillage Research* 115, 16–26.

Piotrowska, A., Wilczewski, E. (2012): Effects of catch crops cultivated for green manure and mineral nitrogen fertilization on soil enzyme activities and chemical properties. *Geoderma* 189, 72–80.

Pompelli, M. F., Martins, S. C. V., Antunes, W. C., Chaves, A. R. M., DaMatta, F. M. (2010): Photosynthesis and photoprotection in coffee leaves is affected by nitrogen and light availabilities in winter conditions. *Journal of Plant Physiology* 167, 1052–1060.

Power, J. F., Biederbeck, V. O. (1991): Role of cover crops in integrated crop production systems, in Hargrove, W. L.: Cover crops for clean water. SWCS, Ankeny, IA, pp. 167–174.

R Development Core Team (2010): R: A language and environment for statistical computing.

Redel, Y. D., Rubio, R., Rouanet, J. L., Borie, F. (2007): Phosphorus bioavailability affected by tillage and crop rotation on a Chilean volcanic derived Ultisol. *Geoderma* 139, 388–396.

References

Roberts, T. L., Stewart, J. W. B., Bettany, J. R. (1985): The Influence of Topography on the Distribution of Organic and Inorganic Soil Phosphorus Across a Narrow Environmental Gradient. *Canadian Journal of Soil Science* 65, 651–665.

Rochette, P., Angers, D. A., Chantigny, M. H., Bertrand, N. (2008): Nitrous oxide emissions respond differently to no-till in a loam and a heavy clay soil. *Soil Science Society of America Journal* 72, 1363–1369.

Saenger, A., Geisseler, D., Ludwig, B. (2011): Effects of moisture and temperature on greenhouse gas emissions and C and N leaching losses in soil treated with biogas slurry. *Biology and Fertility of Soils* 47, 249–259.

Sainju, U. M., Senwo, Z. N., Nyakatawa, E. Z., Tazisong, I. A., Reddy, K. C. (2008): Soil carbon and nitrogen sequestration as affected by long-term tillage, cropping systems, and nitrogen fertilizer sources. *Agriculture Ecosystems & Environment* 127, 234–240.

Sainju, U. M., Singh, B. P., Whitehead, W. F. (2002): Long-term effects of tillage, cover crops, and nitrogen fertilization on organic carbon and nitrogen concentrations in sandy loam soils in Georgia, USA. *Soil & Tillage Research* 63, 167–179.

Salinas-Garcia, J. R., Hons, F. M., Matocha, J. E., Zuberer, D. A. (1997): Soil carbon and nitrogen dynamics as affected by long-term tillage and nitrogen fertilization. *Biology and Fertility of Soils* 25, 182–188.

SAS Institute (2010): SAS/STAT 9.22 User's Guide, Cary (NC). 8444 pp.

Sathya, S., Pitchai, G., James, I. R. (2009): Effect of soil properties on availability of nitrogen and phosphorus in submerged and upland soil – A review. *Agricultural Reviews* 30, 71–77.

Schmidt, J. P., Buol, S. W., Kamprath, E. J. (1996): Soil phosphorus dynamics during seventeen years of continuous cultivation: Fractionation analyses. *Soil Science Society of America Journal* 60, 1168–1172.

Schmidt, M. W. I., Torn, M. S., Abiven, S., Dittmar, T., Guggenberger, G., Janssens, I. A., Kleber, M., Koegel-Knabner, I., Lehmann, J., Manning, D. A. C., Nannipieri, P., Rasse, D. P., Weiner, S., Trumbore, S. E. (2011): Persistence of soil organic matter as an ecosystem property. *Nature* 478, 49–56.

Schneider, M. P. W., Scheel, T., Mikutta, R., van Hees, P., Kaiser, K., Kalbitz, K. (2010): Sorptive stabilization of organic matter by amorphous Al hydroxide. *Geochimica et Cosmochimica Acta* 74, 1606–1619.

References

Scholz, G., Quinton, J. N., Strauss, P. (2008): Soil erosion from sugar beet in Central Europe in response to climate change induced seasonal precipitation variations. *Catena* 72, 91–105.

Schüller, H. (1969): Die CAL-Methode, eine neue Methode zur Bestimmung des pflanzenverfügbaren Phosphates in Böden. *Zeitschrift für Pflanzenernährung und Bodenkunde* 123, 48–63.

Schwertmann, U. V. W. K. M. (1987): Bodenerosion durch Wasser. Vorhersage des Abtrags und Bewertung von Gegenmaßnahmen. Verlag Eugen Ulmer, Stuttgart (Germany).

Shafqat, M. N., Pierzynski, G. M. (2010): Long-Term Effects of Tillage and Manure Applications on Soil Phosphorus Fractions. *Communications in Soil Science and Plant Aalysis* 41, 1084–1097.

Six, J., Conant, R. T., Paul, E. A., Paustian, K. (2002): Stabilization mechanisms of soil organic matter: Implications for C-saturation of soils. *Plant and Soil* 241, 155–176.

Six, J., Elliott, E. T., Paustian, K. (2000): Soil macroaggregate turnover and microaggregate formation: a mechanism for C sequestration under no-tillage agriculture. *Soil Biology & Biochemistry* 32, 2099–2103.

Soane, B.D., Ball, B.C., Arvidsson, J., Basch, G., Moreno, F., Roger-Estrade, J. (2012): No-till in northern, western and south-western Europe: A review of problems and opportunities for crop production and the environment. *Soil & Tillage Research* 118, 66–87.

Stanford, G., Smith, S. J. (1972): Nitrogen Mineralization Potentials of Soils. *Soil Science Society of America Journal* 36, 465–472, https://www.soils.org/publications/sssaj/abstracts/36/3/NP.

Steele, M. K., Coale, F. J., Hill, R. L. (2012): Winter Annual Cover Crop Impacts on No-Till Soil Physical Properties and Organic Matter. *Soil Science Society of America Journal* 76, 2164–2173.

Strosser, E. (2010): Methods for determination of labile soil organic matter: An overview. *Journal of Agrobiology* 27, 49–60, http://www.degruyter.com/view/j/agro.2010.27.issue-2/s10146-009-0008-x/s10146-009-0008-x.xml.

Fact sheet of Agriculture and natural Resources.

References

Tiessen, H., Moir, J. O. (1993): Characterization of available P by sequential extraction. In: Soil sampling and methods of analysis. Lewis Publishers, Boca Raton.

Tiessen, H., Stewart, J. W. B., Cole, C. V. (1984): Pathways of phosphorus transformations in soils of differing pedogenesis. *Soil Science Society of America Journal* 48, 853–858.

Tisdall, J. M., Oades, J. M. (1982): Organic-Matter and Water-Stable Aggregates in Soils. *Journal of Soil Science* 33, 141–163.

v. Luetzow, M., Koegel-Knabner, I., Ekschmitt, K., Matzner, E., Guggenberger, G., Marschner, B., Flessa, H. (2006): Stabilization of organic matter in temperate soils: mechanisms and their relevance under different soil conditions - a review. *European Journal of Soil Science* 57, 426–445.

v. Luetzow, M., Koegel-Knabner, I., Ekschmitt, K., Flessa, H., Guggenberger, G., Matzner, E., Marschner, B. (2007): SOM fractionation methods: Relevance to functional pools and to stabilization mechanisms. *Soil Biology & Biochemistry* 39, 2183–2207.

Vance, E. D., Brookes, P. C., Jenkinson, D. S. (1987): An Extraction Method for Measuring Soil Microbial Biomass-C. *Soil Biology & Biochemistry* 19, 703–707.

VDLUFA: Standpunkt "Phosphordüngung nach Bodenuntersuchung und Pflanzenbedarf". Kerschberger, M.; Hege, U.; Jungk, A. (eds.). Verband Deutscher Landwirtschaftlicher Untersuchungs- und Forschungsanstalten, Darmstadt (Germany).

Verband der Landwirtschaftskammern (2012): Zwischenfrüchte für Futternutzung und Gründüngung. Hinweise zur artenwahl, Nutzungsmöglichkeiten und Anbauverfahren, http://www.lwk-niedersachsen.de/index.cfm/portal/2/nav/278/article/21012.html.

Villamil, M. B., Bollero, G. A., Darmody, R. G., Simmons, F. W., Bullock, D. G. (2006): No-till corn/soybean systems including winter cover crops: Effects on soil properties. *Soil Science Society of America Journal* 70, 1936–1944.

Vu, D. T., Tang, C., Armstrong, R. D. (2009): Tillage system affects phosphorus form and depth distribution in three contrasting Victorian soils. *Australian Journal of Soil Research* 47, 33–45.

Walthert, L., Graf, U., Kammer, A., Luster, J., Pezzotta, D., Zimmermann, S., Hagedorn, F. (2010): Determination of organic and inorganic carbon, delta C-13, and nitrogen in soils containing carbonates after acid fumigation with HCl. *Journal of Plant Nutrition and Soil Science* 173, 207–216.

Wang, W. J., Smith, C. J., Chen, D. (2003): Towards a standardised procedure for determining the potentially mineralisable nitrogen of soil. *Biology and Fertility of Soils* 37, 362–374.

Watts, D. B., Torbert, H. A., Prior, S. A., Huluka, G. (2010): Long-Term Tillage and Poultry Litter Impacts Soil Carbon and Nitrogen Mineralization and Fertility. *Soil Science Society of America Journal* 74, 1239–1247.

Wright, A. L. (2009): Phosphorus sequestration in soil aggregates after long-term tillage and cropping. *Soil & Tillage Research* 103, 406–411.

Zamuner, E. C., Picone, L. I., Echeverria, H. E. (2008): Organic and inorganic phosphorus in Mollisol soil under different tillage practices. *Soil & Tillage Research* 99, 131–138.

Zhou, X., Chen, C., Wu, H., Xu, Z. (2012): Dynamics of soil extractable carbon and nitrogen under different cover crop residues. *Journal of Soils and Sediments* 12, 844–853.

Zibilske, L. M., Bradford, J. M., Smart, J. R. (2002): Conservation tillage induced changes in organic carbon, total nitrogen and available phosphorus in a semi-arid alkaline subtropical soil. *Soil & Tillage Research* 66, 153–163.

Zorn, W., Schröter, H., Wagner, S. (2011): Ohne Pflug anders düngen, in DLG-Mitteilungen 2011.

Zotarelli, L., Alves, B. J. R., Urquiaga, S., Boddey, R. M., Six, J. (2007): Impact of tillage and crop rotation on light fraction and intra-aggregate soil organic matter in two Oxisols. *Soil & Tillage Research* 95, 196–206.

I want morebooks!

Buy your books fast and straightforward online - at one of the world's fastest growing online book stores! Environmentally sound due to Print-on-Demand technologies.

Buy your books online at
www.get-morebooks.com

Kaufen Sie Ihre Bücher schnell und unkompliziert online – auf einer der am schnellsten wachsenden Buchhandelsplattformen weltweit!
Dank Print-On-Demand umwelt- und ressourcenschonend produziert.

Bücher schneller online kaufen
www.morebooks.de

OmniScriptum Marketing DEU GmbH
Heinrich-Böcking-Str. 6-8
D - 66121 Saarbrücken
Telefax: +49 681 93 81 567-9

info@omniscriptum.com
www.omniscriptum.com

Printed by Books on Demand GmbH, Norderstedt / Germany